Kezia Balena

Bioremediation of cadmium under laboratory conditions

Kezia Balena

Bioremediation of cadmium under laboratory conditions

Evaluation of the potential of yeasts isolated from the fermentation of regional cachaça for bioremediation

ScienciaScripts

Imprint

Any brand names and product names mentioned in this book are subject to trademark, brand or patent protection and are trademarks or registered trademarks of their respective holders. The use of brand names, product names, common names, trade names, product descriptions etc. even without a particular marking in this work is in no way to be construed to mean that such names may be regarded as unrestricted in respect of trademark and brand protection legislation and could thus be used by anyone.

Cover image: www.ingimage.com

This book is a translation from the original published under ISBN 978-613-9-68281-2.

Publisher:
Sciencia Scripts
is a trademark of
Dodo Books Indian Ocean Ltd. and OmniScriptum S.R.L publishing group

120 High Road, East Finchley, London, N2 9ED, United Kingdom
Str. Armeneasca 28/1, office 1, Chisinau MD-2012, Republic of Moldova, Europe
Printed at: see last page
ISBN: 978-620-8-20600-0

SUMMARY

"Evaluation of the potential of yeasts isolated from the fermentation of cachaça and regional ecological niches for ecological niches for cadmium bioremediation under laboratory conditions"

Kezia Pereira de Oliveira Balena

ACKNOWLEDGMENTS

To my beloved brother Jetro Pereira de Oliveira, who has often been more than a father, supporting me financially even in the face of so many difficulties, and with words of great wisdom, because he has walked this same path and knows the thorns that are found in them, to him who was certainly the key to achieving this victory.

To my sister Sara and brother-in-law Ricardo, who were always willing to listen to me and give me affection, where distance was no barrier to the great affection that unites us.

To my beloved husband Devair Balena Jûnior for inspiring me to believe in my dreams and helping me to become a better person every day.

To the CDTN for providing the facilities and equipment to carry out the work. Many thanks to all the staff and members of the CDTN who helped me many times.

I thank God for this victory, for always showing me the best path to follow, to him who answered my prayers where my only request was that I could have more humility and wisdom every day, for my faith and trust, because if it weren't for his divine power, so many achievements would certainly be impossible.

SUMMARY

Evaluation of the potential of yeasts isolated from cachaça fermentation and regional ecological niches for cadmium bioremediation under laboratory conditions

Kezia Pereira de Oliveira Balena

SUMMARY

Unlike organic pollutants, heavy metals cannot be chemically degraded. Therefore, chemical and physical remediation processes are used to remove them before they are discharged into water bodies. Most of these processes are effective, but they can be economically unfeasible or aggravate the problem, as large quantities of chemical reagents may be required. The addition of chemical reagents can remove the metal, but creates a new problem with the disposal of the chemicals. Bioremediation, using microorganisms as biosorbent material, has been successfully applied in several cases.

Yeast cells can accumulate metal ions on the cell surface through the process of adsorption, which is a passive process, or they can actively absorb the metal, which is accumulated inside the cell.

In this work, several biological, physiological and chemical parameters were studied, which are important for selecting yeasts that can be used to capture cadmium from aqueous media.

Fungi, yeasts and bacteria, when exposed to adverse situations, show an adaptive response that enables the cell to survive the stress. Normally, in these situations, the carbohydrate trehalose is synthesized. We observed that the presence of cadmium chloride induces the synthesis of trehalose; however, the accumulation of this carbohydrate does not play a fundamental protective role under the conditions studied.

Yeast can incorporate cadmium in different chemical forms. The use of different cadmium chemical compounds, such as chloride, oxide, acetate, nitrate and sulfate, does not provide differentiated incorporation of the metal by the yeast.

The yeast strains isolated from the fermentation of cachaça and from the region showed greater tolerance to high concentrations of cadmium when compared to laboratory strains. All *Saccharomyces cerevisiae* (with the exception of the laboratory strains) tolerate a concentration of 50 ppm cadmium chloride for their growth, while the yeast *Starmerela meliponinorum* grows at 100 ppm. The laboratory strain *Saccharomyces cerevisiae* only tolerates 10 ppm of cadmium chloride for its growth. High concentrations of cadmium chloride inhibit cell growth in all the yeast strains studied.

Cadmium incorporation by yeast cells depends on the initial concentration of the metal in the medium and the length of time the cells are exposed to the metal. When comparing short periods of exposure to cadmium, 3 hours, with long periods, 24 hours, cadmium incorporation by the cells was greater when the cells remained in contact with the metal for longer.

For living cells, cadmium incorporation depends on the phase of growth the cell is in. Cells in the stationary phase take up less cadmium than cells in the logarithmic phase of growth.

The increase in mass does not favor the incorporation of the metal. On the other hand, yeasts killed by autoclaving capture more cadmium than live yeasts.

Several yeast strains were used in this work. We used nine *Saccharomyces cerevisiae* strains isolated from the fermentation of cachaça, from stills in the state of Minas Gerais, and six yeast strains representing different genera and species of the *Saccharomyces* genus, isolated from regional ecological niches. We did not find that one specific strain had much higher cadmium incorporation than another. The results are similar, indicating that the ability of live yeasts to incorporate cadmium is not closely linked to the genus or species used.

Our studies showed that yeasts isolated from the fermentation of cachaça and those selected in the region showed greater incorporation of cadmium when compared to laboratory strains *of Saccharomyces cerevisiae*.

This work indicates that it is possible to use alembic yeast biomass to incorporate cadmium from contaminated effluents, without the need to isolate the strains that compose it, using only the biosorbent and bioaccumulative properties of the total yeast biomass.

1 INTRODUCTION

In recent decades, development and rapid industrialization have brought serious environmental problems. Industrial waste containing heavy metals is dumped in a disorderly manner and constitutes a major risk to health and the environment.

Heavy metals are chemical elements that have a specific weight of more than 5 g.cm^{-3} , and are considered "trace elements" because they are naturally found in a few parts per million (ppm) (MATTIAZZO-PREZOTTO, 1994). They differ from other toxic agents in that they are neither synthesized nor degraded by humans. Industrial activity, aimed at producing new compounds, has significantly reduced the permanence of these metals in ores, altering the distribution of these elements on the planet.

The action of heavy metals on human health is very diverse. Among the most dangerous metals are mercury, cadmium (found in cell phone batteries), chromium and lead.

The term "heavy metal" is commonly used to refer to metals classified as pollutants, encompassing a very heterogeneous group of metals, semi-metals and even non-metals such as selenium. The list of heavy metals most frequently includes the following elements: copper, iron, manganese, molybdenum, zinc, cobalt, nickel, vanadium, aluminum, silver, cadmium, chromium, mercury and lead (CETESB, 2001).

Industrial activities have introduced heavy metals into water in much greater quantities than would be natural, causing pollution. To get an idea of this, it is enough to remember that heavy metals are part of the discharges of large industries in all countries of the world.

In Brazil, it is estimated that of the 2.9 million tons of hazardous industrial waste generated annually, only 600,000 tons receive adequate treatment, according to an estimate by the Brazilian Association of Special Waste Treatment, Recovery and Disposal Companies - ABETRE (WHO, 1992). The remaining 78% is improperly disposed of in garbage dumps, without any kind of treatment (WHO, 1992).

Growing industrialization in recent decades has exposed animals and plants to many potentially toxic chemicals. Heavy metals or trace metals stand out as the most dangerous because, even at trace levels, they can cause irreversible eco-biological damage (MIRANDA, 1993).

All forms of life are affected by the presence of metals, depending on the dose and chemical form. Many metals are essential for the growth of biological organisms, from bacteria to humans, but they are required in trace concentrations; above these levels, they can damage biological systems. As a result of the phenomenon of bioaccumulation, sub-toxic quantities present in the environment can reach risk levels in the final links of the trophic chain (VOLESKY, 1990a).

1.1 Cadmium

Along with copper and zinc, cadmium belongs to group IIb on the Periodic Table. It is a relatively rare element and is not found in nature in its pure state. It is mainly associated with sulphides in zinc, lead and copper ores. It was first purified in 1817, but its commercial production only became important at the beginning of the last century. When heated to high temperatures it emits highly toxic cadmium fumes. Cadmium is slowly oxidized by moisture in the air to cadmium oxide (ILO, 1998) and is also a great neutron absorber (HSDB, 2000).

In 2001, cadmium was ranked 70th on the "Most Hazardous Substances" list of CERCLA (Comprehensive Environmental Response, Compensation, and Liability Act), together with the EPA (Environmental Protection Agency) and ATSDR (Agency for Toxic Substances and Disease Registry), where substances are classified according to their toxicity, health risk potential and exposure to living organisms.

The main natural source of cadmium in the atmosphere is volcanic activity. Cadmium emissions occur both during eruptions and during periods of low volcanic activity. This source, although very difficult to quantify, has been estimated to be responsible for the emission of 820 tons of the metal per year. Deep-sea volcanic activity is also an important part of the cadmium cycle, but it has not yet been possible to quantify it (WHO, 1992).

Cadmium has been increasingly used in industry, as it is used to produce pigments for paints, plastics, batteries and electrolytic deposits. Worldwide, its annual production is 20,000 tons, most of which is destined for the electrochemical deposit industry. It is interesting to note that 50% of the annual cadmium is consumed in the USA, and that the environment can be affected at a distance of 100 km in order of magnitude (U.S. GEOLOGICAL SERVEY, 2005).

Brazil has a very diversified mineral production and stands out in the global context as a major mineral exporter. The country is self-sufficient in most mineral products. External dependence lies particularly in metallurgical coal, potash and raw materials for the metallurgy of non-ferrous metals such as copper, zinc and, consequently, cadmium. Cadmium does not appear among Brazil's main mineral reserves, but zinc does appear in a small proportion (U.S. GEOLOGICAL SERVEY, 2005).

Cadmium has limited use. Its main applications fall into five main categories (WHO, 1992):

- In steel and iron coating;

- As a stabilizer for polyvinyl chloride (PVC);

- In pigments for plastics and glass;

- Nickel-cadmium batteries;

- Alloys.

In addition to these uses, other uses are cited in the literature:

- Fungicide (cadmium chloride) (ILO, 1998);

- Additive in the textile industry (ILO, 1998);

- Photographic film production (ILO, 1998);

- Manufacture of special mirrors (ILO, 1998);

- Electronic vacuum tube covers (on electrodes and cadmium vapor lamps) (HSDB, 2000; ILO, 1998);

- Semiconductors (ILO, 1998);

- Glass and glazed ceramics (ILO, 1998);

- Aluminum welding (MEDITEXT, 2000);

- Fire protection system (MEDITEXT, 2000);

- Television (HSDB, 2000);

- Neutron absorber in nuclear reactors (HSDB, 2000);

- Amalgam in dental treatment (1Cd: 4Hg) (HSDB, 2000);

- Power transmission wires (MEDITEXT, 2000).

Of all the cadmium used, 75% is used in Ni-Cd batteries, and the remaining 25% is distributed as follows: pigments 13%; coating and deposition 7%; stabilizers for plastics 4%, non-ferrous alloys and other uses 1%. Cadmium consumption varies from country to country, depending on environmental restrictions, industrial development, natural sources and commercial levels (U.S. GEOLOGICAL SURVEY, 2005).

Cadmium toxicity was discovered more than a century ago and its use has expanded significantly (LASKEY & REHNBERG, 1984). Cadmium is considered a probable carcinogenic or carcinogen-inducing element in humans (VOLESKY, 1990c; ADAMIS et al, 2003).

The main problem with cadmium in humans is that the human body rarely excretes as much cadmium as it absorbs. Cadmium tends to accumulate in the human body (30 mg on average in the American male), with 33% in the kidneys and 14% in the liver (LÓPEZ-ALONSO et al, 2000). Cadmium absorbed by humans (and other animals) has a cumulative effect. It is concentrated in the urine and blood (ROELS et al, 1999; ARANHA et al, 1994), with accumulation in the liver and kidneys (FRIBERG et al, 1974; IKEDA et al, 2000; ROELS et al, 1999; ARANHA et al, 1994). Clinically, acute and chronic cases of cadmium toxicity have increased in humans (RAGAN & MAST, 1990). Contact occurs primarily via the inhalation and digestive routes. Inhalation of cadmium dust and fumes comes from industries. It is also present in tobacco and smokers are therefore subject to greater exposure. Acute inhalation results in pulmonary edema and irritation of the respiratory tract, while chronic inhalation causes emphysematous and fibrotic changes in the lung tissue, as well as damage to the kidney tubules (ZAVON & MEADOW, 1970; FRIBERG et al, 1974).

The best-known case of cadmium poisoning via food occurred on the banks of the *Jintsu* River in the *Funchu-Machi* region of Japan after World War II, when a large number of rice farmers and fishermen suffered rheumatic pains accompanied by bone deformities and kidney disorders. It was later determined that this epidemic was due to cadmium poisoning, which was caused by the consumption of rice contaminated by irrigation water from the effluent of a zinc-lead processing industry. The disease became known in medical science as *Itai-Itai* (ATSDR, 1997; MEDITEXT, 2000; WHO, 1992).

Food is therefore the main source of contamination, with around 70% of exposure occurring orally (GROTEN & BLADEREN, 1994). This contamination can occur during the use of contaminated water for irrigation or through contaminated soil (MIDIO & MARTINS, 2000; JARUP et al, 1998; PRASAD, 1995; TYLER, 1990).

When assimilated, cadmium is distributed throughout the body, being found in blood cells either bound to plasma serum proteins such as albumin and other glycoproteins or bound to metalloproteins (metallothioneins) produced by the liver (MATTIAZZO-PREZOTTO, 1994).

In general, cadmium chloride and cadmium acetate are the most absorbed and more toxic than other compounds (MASON, 2005). As a safety measure, in the USA the safe level of cadmium in drinking water is 10 ppb (part per billion). In Brazil, the concentration permitted by the National Environmental Council (CONAMA) varies from 40 ppb to 1 ppb, depending on the class of water. For example: for class 1 water, the permitted cadmium limit is 1 ppb. Class 1 water is water that, after disinfection, can be used for domestic supply, contact recreation, irrigation of fruit and vegetables and breeding of species intended for human consumption. For class 3 water, i.e. water that, after conventional treatment, can be used for domestic supply, the permitted limit is 10 ppb - Resolution Nº 357-CONAMA (CONAMA, 2005).

With regard to the incorporation of cadmium by yeasts, it is known that microorganisms have developed various mechanisms to rid themselves of the toxic effects of heavy metals or organic compounds. The detoxification of metals in microorganisms is carried out by various mechanisms including: regulation of uptake, transformation into less toxic species and immobilization inside cells. Inside cells, the major molecules involved in metal sequestration are the tripeptide glutathione (GSH; L- -Glu-Cis-Gli), and the phytochelatins or cadistins (-Glu-Cis)n Gli), and low molecular weight proteins rich in cysteines called metallothioneins. The concentration of GSH in *Saccharomyces cerevisiae* can reach 1% of the cell's dry weight. GSH acts as a kind of endogenous nitrogen and sulphur store and plays an important role in protecting against metals. The defensive action of GSH against heavy metal toxicity is based on the observation of its accumulation in response to the presence of metal ions (GADD & GHARIEB, 2004).

Yeast cells have developed a variety of detoxification mechanisms to eliminate toxic compounds. The four main mechanisms described for metal resistance in the yeast *Saccharomyces cerevisiae* are as follows (ERASO et al, 2004):

1- Yeasts have a mechanism that decreases the influx of metals into the cell. This mechanism depends on the repression of the metal transporter gene or the induction of proteolysis of this transporter.

2- Metal binding and the formation of complexes such as metallothioneins (MT). These molecules neutralize the toxic effects of free metals by binding them to cells. The yeast *Saccharomyces cerevisiae* produces metallothioneins capable of binding copper and cadmium. These MTs are encoded by the same CUPI gene, whose transcription is regulated differently by copper or cadmium.

3- Compartmentalization in vacuoles. The vacuolar membrane protein Ycfl (yeast cadmium factor) plays an important role in cadmium tolerance in *Saccharomyces cerevisiae*. This protein is a member of the ABC (ATP Binding Cassette) transporter superfamily and has high homology with the human Mrpl (human multidrug resistance-associated protein). Ycfl is a vacuolar protein, a pump that conjugates glutathione and sulphur and transports the Cd-GSH complex to the vacuole. Cells overexpressing the YCFl gene show resistance to cadmium.

4- Active export system that limits the intracellular concentration of the metal, like the efflux system described for bacteria. This mechanism is not widely described for *Saccharomyces cerevisiae* or other fungi (SHIEAISHI et al, 2000).

Apparently, the most common mechanism in yeast is the "glutathione S-conjugates" (GS-X), part of the ATP-binding cassette (ABC) family of transporters. GS-X pumps catalyze the transport of glutathione conjugates to the vacuole. Through this transporter cadmium is complexed to glutathione, removed from the cytoplasm and transported to the vacuole (ERASO et al, 2004).

These GS-X transporters are involved in the removal from the cytoplasm of a range of S-conjugated compounds. To date, two GS-X transporters have been molecularly identified: MRPl, human multidrug resistance associated protein, and YCF, yeast cadmium factor. These two proteins have 43% identity and 63% homology (LI et al, 1997).

In *Saccharomyces cerevisiae*, the regulation of the adaptive response to heavy metals and oxidants occurs mainly at the transcriptional level. The response of the yeast *Saccharomyces cerevisiae* to cadmium is typical, i.e. there is an overlap between the response to various stresses, which may reflect the production of signals common to these stresses or common regulatory components or both. For example, several heat shock proteins are transcribed in response to cadmium. The tripeptide glutathione plays a central role in protection against oxidative stress and cadmium toxicity. The proteomic approach reveals that up to 54 proteins have increased expression in the presence of cadmium, many of which play a key role in oxidative stress, corroborating the hypothesis that exposure to heavy metals results in oxidative stress. At the same time, during exposure of *Saccharomyces cerevisiae* yeast cells to cadmium, there is a substantial reduction in the level of 43 other proteins (JAMIESON, 2002).

Post-transcriptional regulation is also involved in the response to heavy metals. A typical example is the synthesis of peptides called phytochelatins (PCs) after exposure to heavy metals. PCs are synthesized from GSH in a reaction catalyzed by PC synthases, PCs have a general structure (Glu-Cys)n-Xaa, containing 2-11 Glu-Cis repeats. PCs chelate heavy metals with high affinity and facilitate the vacuolar sequestration of these metals (OLENA et al, 2001).

1.2 Accumulation of metals by yeasts

Fungi and yeasts are able to remove heavy metals from the external environment through mechanisms that can be physical-chemical, such as adsorption, or dependent on metabolic activity, such as transport. Some physicochemical interactions can be indirectly dependent on metabolism via the synthesis of particular constituents of the cell or metabolites that can act as efficient chelators or the creation of a particular microenvironment close to the cell that facilitates deposition or precipitation. Thus, microbial biomass, living or dead, is capable of accumulating metal (GADD, 1990).

Yeasts can act as accumulators of toxic substances, including heavy metals, for which they have the capacity to capture (BRADY et al, 1994; VOLESKY, 1990a).

The yeast cell, with its complex cell wall, represents an additional adsorption site compared to cells without a cell wall. The cell wall is considered to protect the plasma membrane and the cell as a whole (GADD, 1990).

Numerous chemical parameters must be considered in order for a metal ion to be adsorbed by a living cell. These include: metal charge, ionic radius, the metal's preference for organic ligands and the availability and concentration of the metal (WOOD & WANG, 1983).

The ability of microorganisms to accumulate heavy metals generally involves two phases: rapid binding to the cell surface, independent of metabolism, followed by metabolism-dependent intracellular accumulation, with energy expenditure. In metabolism-independent accumulation, the cations can be deposited by adsorption, inorganic precipitation or adsorption to anionic groups present in the cell wall (BRADY & DUNCAN, 1994; VOLESKY, 1990b; VOLESKY & MAY-PHILLIPS, 1995). Thus, biosorption is the ability of the cell to passively sequester the metal by different physico-chemical mechanisms. This ability is dependent on factors external to the microorganism, as well as the metal, its ionic form in solution, and the active retention site responsible for sequestering the metal. An important characteristic of biosorption is that it can occur even when the cell is no longer metabolically active, i.e. when it is dead (VOLESKY, 1990a). The biosorption of heavy metals by dead biomass has been used as an alternative to existing removal technologies used in wastewater treatment. The use of dead biomass circumvents the problem of metal ion toxicity to cells (MATIS & ZOUBOULIS, 1994).

In metabolism-dependent accumulation, the metal is transported by transport proteins across the cell membrane and accumulates in the cytosol bound to metallothionein (BRADY & DUNCAN, 1994). Metallothionein is involved in metal storage, detoxification, development, differentiation, control of cell metabolism, protection against free radicals and response to ultraviolet radiation (BRADY & DUNCAN, 1994; VOLESKY, 1990c).

There are different mechanisms of action of complexing agents in biological systems. Microorganisms can complex metals in solutions, in extracellular polymers on the cell surface or in intracellular compartments (WOOD & WANG, 1983).

The types of complexation that cells can establish with metals are: complexation and efflux, biosorption on the cell surface, complexation via specific mechanisms, extracellular release of complexing agents, complexation and storage in intracellular compartments (BIRCH & BACHOFEN, 1990).

Biosorption and bioaccumulation are processes that result in the reduction of toxicity. Depending on their size or charge, the complexes formed cannot pass through the cell membrane and, because they are insoluble, they precipitate. On the other hand, low molecular weight complexes can be formed and enter the cell by diffusion, which can then be exported back into the environment or stored in intracellular compartments (MACASKIE et al, 1987; PETTERSON et al, 1985; TYNECKA et al, 1981).

These different mechanisms are hypotheses found in the literature to explain the differences in toxicity observed in certain metals and may play a significant role in environmental impacts (BABICH & STOTZKY, 1980).

A metal complexed with an organism generally has a different level of toxicity than that presented in its free form (BABICH & STOTZKY, 1983; SPOSITO, 1983).

There are considerable differences in the production of complexing compounds depending on the growth phase of the organism; cells in the exponential phase of growth produce different complexing materials from those formed in the stationary phase (GADD, 1990).

Microbiological chelating agents do not necessarily have to be released into the environment, but can remain in intracellular compartments and on the outer surface of the cell, for example in the form of polysaccharides or polymers. They can be part of the cell wall or in capsule form, acting as efficient biosorbent compounds (BEVERIDGE, 1989; KAPLAN et al, 1987; MACASKIE et al, 1987).

In some cases, adsorption is followed by internalization of another complex via an active process, as is the case with essential metals and some toxic metals, or by a passive diffusion process, which is believed to be the adsorption route for most toxic metals (GADD, 1990).

1.3 Trehalose

Exposure of yeasts to environments polluted with heavy metals is a very harmful stimulus. To survive in this new situation, the cell must adapt and survive. If the adaptation is not effective, the cell will die. One of the mechanisms by which yeasts become more resistant and adapt to a harmful situation is the accumulation of intracellular trehalose. We therefore chose to determine trehalose as one of the markers of stress. Trehalose is a non-reducing disaccharide made up of two glucose units. Originally, trehalose, together with glycogen, were considered reserve carbohydrates for microorganisms, but in the last two decades, several authors have shown that trehalose has, in addition to its reserve function, the function of protecting cells during stress processes, such as high temperatures, osmotic shock, exposure to ethanol and dehydration (VAN LAERE, 1989). Trehalose is the least reactive and most stable carbohydrate in nature (PANEK, 1995). This sugar has been implicated in survival under various stressful conditions, acting as a membrane protector (CROWE et al, 1984; LESLIE et al, 1995), as well as a compatible solute or as a reserve carbohydrate. Cells in stationary phase and those growing on non-fermentable carbon sources are naturally more resistant to stress and have a higher level of trehalose (VAN DIJCK et al, 1995). In the yeast *Saccharomyces cerevisiae*, trehalose can make up more than 23% of the cell's dry weight, depending on growth conditions and the stage of the cell cycle (THEVELEIN, 1984). The accumulation of trehalose that occurs in the presence of a stressful agent is sometimes dependent on the time of exposure.

The protective role of trehalose can be demonstrated when the temperature of the incubation medium is increased from 30°C to 40°C. This increase in temperature promotes the accumulation of trehalose in yeasts. Subsequently, if these yeasts, with a high trehalose content, are transferred to a temperature of 55^0 C (lethal temperature) it is observed that a large proportion of the cells survive what was previously a lethal temperature (NEVES et al, 1991).

There is a strong association between high levels of trehalose and the resistance of yeasts to freezing, a fact that is of great importance in baker's yeast that is marketed frozen (HINO et al, 1990).

The survival of cells depends on their ability to monitor the extracellular environment and quickly respond with changes that allow them to survive in the new situation. Normally, changes in the physical or chemical

12

conditions of the cell's environment lead to an immediate block in cell growth and trigger a series of cellular responses that are essential for survival. These responses include the immediate production of compounds called compatible solutes such as trehalose. The term compatible solute is due to the fact that even when accumulated in high concentrations, the solute does not interfere with the cell's normal metabolism.

The synthesis of trehalose is the result of the action of pathways that detect stress and signal metabolic pathways that will induce the expression of genes, the products of which can confer protection to cells.

In yeast, trehalose is formed in the cytosol by a two-step reaction (CABIB & LELOIR, 1958). This reaction is carried out by a complex of three subunits: a catalytic subunit which is the trehalose-6-phosphate synthase encoded by the *TPS1* gene (BELL et al, 1998), a subunit which is the trehalose-6-phosphate phosphatase encoded by the *TPS2* gene (DE VIRGILIO et al, 1993), the largest subunit encoded redundantly by 2 genes, *TSL1* (long-chain trehalose synthase) and *TPS3* with regulatory activity (VUORIO et al, 1993). The two enzymes involved in this process are dependent on Mg^2 + (ELLIOT et al, 1996). The *TPS3* and *TSL1* subunits synthesize proteins responsible for stabilizing the complex (BELL et al, 1998). Threose-6-phosphate synthase can function independently of the complex (BELL et al, 1998). The genes encoding the components of the complex are induced by stress (DE VIRGILIO et al, 1993; WINDERICKX et al,1996).

The trehalose synthase/phosphatase complex is also involved in regulating the influx of glucose into the cell. Deletion of the *TPS1* gene causes a series of regulatory problems such as: activation of fructose-1,6-biphosphatase, inactivation of plasma membrane H^+ -ATPase, inactivation of the potassium uptake system, loss of pyruvate decarboxylase activity, decrease in the content of trehalose in non-fermentable carbon sources, hyperaccumulation of bisphosphate sugars, gradual decrease in the polyphosphate pool, intracellular acidification, etc (VAN AELST et al, 1993).

The breakdown of trehalose into 2 glucose molecules is carried out by acid or vacuolar trehalase, encoded by the *ATH1* gene, and by neutral or cytosolic trehalase, encoded by the *NTH1* gene. Currently, with the new classification proposed, the name would be: neutral or cytosolic trehalase and acid or extra-cellular trehalase (PARROU et al, 2005). Acid trehalase has optimal activity at pH 4.0 and 5.0 (LONDESBOROUGH & VUORINO, 1984) and its activity is only detected when the cell enters the respiratory and stationary phases or when the cell grows on a non-fermentable substrate such as ethanol and glycerol (SAN MIGUEL & ARGUELLES, 1994). Neutral trehalase activity occurs at pH 6.7 and 7.0 (LONDESBOROUGH & VUORINO, 1984) and its activity is regulated by cAMP-dependent phosphorylation mechanisms (COUTINHO et al, 1992). The activity of this enzyme is high in the exponential phase, when trehalose levels are low and in the presence of fermentable sugars, such as glucose (NEVES & FRANÇOIS, 1992).

The accumulation of trehalose is related to glucose, nitrogen and phosphate fasting (LILLIE & PRINGLE, 1980) as well as protection against dehydration and desiccation stress (D'AMORE et al, 1991; GADD et al, 1987), freezing (LEWIS et al, 1995), thermal stress (DE VIRGILLO et al, 1994; LEWS et al, 1995; NEVES et al, 1991; NEVES & FRANÇOIS, 1992), osmotic (HOUNSA et al, 1998) and chemical products such as ethanol (LUCERO et al, 2000; MANSURE et al, 1994; RIBEIRO et al, 1999), heavy metals (ATTFIELD, 1987), oxygen radicals (BENAROUDJ et al, 2001) and hydrostatic pressure (FERNANDES et al, 2001). The trehalose content in yeast is probably one of the most important factors affecting yeast resistance during lyophilization and subsequent rehydration. In these cases, the increase in trehalose levels is closely related to the increase in viability, although in some cases, such as osmotic stress, ethanol tolerance and pH extremes, glycerol seems to be more closely involved (HOUNSA et al, 1998; SIDERIUS et al, 2000).

In general, cells with a high trehalose content are more resistant to various harmful situations. The presence of heavy metals is one of these factors which, when present in the environment, the cell must monitor and trigger protective mechanisms (ATTFIELD, 1987).

1.4 Bioremediation

Anthropogenic activities result in pollution of the urban and rural environment, creating vast areas that are unsafe or uninhabitable, not to mention major environmental disasters such as Chernobyl and the Exxon Valdez oil spill. So remediation is big business all over the world. The USA has around 12,000 sites listed as contaminated and Europe has around 400,000 of these sites. It is estimated that there are still thousands of unofficial sites (WATANABA, 2001). In 1998, the remediation market was worth around 15-18 billion dollars (WATANABA, 2001). Most remediation is of groundwater and soil. Many areas are contaminated with a combination of heavy metals and organic compounds, and many of these sites also contain radionuclides. Conventional remediation involves removing the contaminant from the site and disposing of it elsewhere. For example, contaminated soil is excavated and then replaced with new soil. Contaminated water is pumped out, the pollutant removed by methods such as filtration. Clearly these technologies do not remediate the contamination, they simply remove it from one site to another or transform it from one state to another (e.g. from liquid to gas). In this way, remediation projects can take decades to complete, are very difficult and can leave residual contamination (WATANABA, 2001). It is in this problematic context that the technology of remediation with biological materials (microorganisms such as bacteria, fungi, yeasts, algae or plants) emerges, i.e. bioremediation. Bioremediation can be carried out *in situ* or *ex situ*. Bioremediation of organic compounds can be faster than conventional techniques and, above all, can leave no residual contamination. Microorganisms and plants can be used to remove heavy metals from water or soil, but they must usually be disposed of by incineration or cement (in the case of radionuclides). The US EPA (Environmetal Protection Agency) estimates that phytoremediation can save 50-80% of the cost of conventional technologies (WATANABA, 2001).

Bioremediation is a technology for the removal and recovery of heavy metals from contaminated areas. COSSICH et al 2000 define bioremediation as "a process in which solids of plant origin or microorganisms are used to retain, remove or recover heavy metals from a liquid environment". Bioremediation has received a great deal of attention, both as a scientific novelty and for its possible application in industry. However, as a technology, bioremediation is not new. For example, the use of composting in agriculture and residential sewage treatment are techniques based on the use of microorganisms to catalyze chemical transformations. These environmental management techniques have been applied by humanity since the dawn of civilization, and their use predates the discovery of the existence of microorganisms. The Romans used bioleaching to recover copper in the mines of Spain, at a time when their empire extended to this region. In his time, the scientist Paracelsus (1493-1541) described bioleaching, i.e. the recovery of copper from mines (OLSON et al, 2003). The most "modern" use of bioremediation dates back 100 years, with the opening of the first biological sewage treatment plant in Sussex, England, in 1891.

Currently, bioremediation is used commercially to remove a limited range of contaminants, mainly the hydrocarbons found in gasoline. However, microorganisms have the ability to biodegrade practically all organic and many inorganic contaminants (NATIONAL RESEARCH COUNCIL, 1993).

Nowadays, the US State Department has a specific program called NABIR (Natural and Accelerated Bioremediation Research) that explores bioremediation potential in order to find solutions for environments contaminated by metals or radionuclides.

New is the term bioremediation, which appeared in scientific literature in 1987 (PALMISANO & HAZEN, 2003).

Bioremediation is the use of microorganisms to reduce, eliminate or contain harmful and/or radioactive compounds to safe levels in the environment. Bioremediation of organic compounds involves transforming them into less toxic or non-toxic products, such as CO_2. Bioremediation of metals and radionuclides involves removing them from the aqueous phase to reduce the risk to humans and the environment. Microorganisms can directly transform metals or radionuclides by changing their oxidized state to a reduced form that allows them to be immobilized. Microorganisms can also indirectly immobilize metals and radionuclides by reducing them to inorganic ions, which in turn chemically reduces the contaminants to less mobile forms. The long-term stability of these contaminants is unknown. Other mechanisms by which microorganisms can influence mobility include: pH alteration, oxidation, and complexation (PALMISANO & HAZEN, 2003).

Microorganisms including bacteria, algae, filamentous fungi and yeasts are efficient bioremediators.

The ability of certain fungi and yeasts to concentrate metals has been used to extract metal species from aqueous media. There is therefore great interest in using biomass to biosorb and detoxify industrial effluents by removing metallic components from them.

The relationship between various microorganisms and heavy metals is well documented. The study of interactions between yeasts and heavy metals is of great scientific interest. Among fungi, yeasts, especially the yeast *Saccharomyces cerevisiae*, are the most scientifically exploited, due to the fact that they are organisms that are easy to manipulate genetically, have a rapid life cycle and grow in relatively inexpensive media, and thus serve as an excellent model for the study of many important problems in the biology of eukaryotes (BROCK et al, 1984). Yeasts are also known to accumulate large quantities of heavy metals from aqueous media (GADD, 1986).

Conventional techniques used to remove metals from industrial effluents include chemical precipitation, ion exchange (KEFALA et al, 1999; KAPOOR & VIRARAGHAVAN, 1995), reverse osmosis (KAPOOR & VIRARAGHAVAN, 1995), chemical oxidation and reduction, filtration, electrochemical treatment, evaporation or solvent extraction (WAIHUNG et al, 1999), however, the cost of these processes is very high (VEGLIO & BEOLCHINI, 1997). Methods such as chemical precipitation and solvent extraction employ a large quantity of chemical reagents which sometimes solve the metal problem, but create the problem of the waste of the chemicals used in these methods.

Bioremediation, i.e. the removal of heavy metals from the environment by microorganisms, is carried out through physical-chemical mechanisms, such as adsorption, or metabolic activity-dependent mechanisms, such as transport. Some physicochemical interactions may be indirectly linked to metabolism, especially via the synthesis of cellular constituents or metabolites that can act as efficient metal chelators, such as glutathione, phytochelatins and metallothioneins (VOLESKY & MAY-PHILLIPS, 1995).

Other processes by which microorganisms accumulate metals are sequestration, transport, precipitation and oxidation-reduction reactions. The accumulation of metals by passive means, adsorption and/or complexation is called biosorption. However, if this accumulation depends on the metabolic activity of the microorganism, then it is bioaccumulation.

For yeasts, the toxicity of heavy metals modifies the biological activity of cellular components such as nucleic acids, enzymes, amino acids and lipids (BRENNAN & SCHIESTL, 1996; ROMANDINI et al, 1992). metal ions can have toxic effects when they accumulate at extremely high levels. Excess Fe and Cu can generate reactive oxygen species that degrade macromolecules such as DNA, proteins and lipids. Metals can also inhibit biochemical processes by competing with other ions for the active site of enzymes, intracellular transporters and other biologically important ligands (ALBERTINI, 1999; ASSMANN et al, 1996; SILVA, 2001), among others. Bioaccumulation is the retention and concentration of a substance within the organism. In bioaccumulation, the solute is transported outside the cell through the cell membrane and sequestered in the cytoplasm. Biosorption is the association of the substance with the cell surface. Sorption does not require active cellular metabolism (PALMISANO & HAZEN, 2003).

In biosorption, microorganisms sequester the metal through surface bonds, while in the bioaccumulation process, metals are concentrated through a combination of surface reactions such as precipitation and the formation of intra- and extracellular complexes (VOLESKY, 1990c). However, there are significant practical limitations to systems that employ bioaccumulation, such as the inhibition of cell growth when the concentration of metal ions becomes too high or the high toxicity of waters such as pH extremes and high salt concentrations. The active process of bioaccumulation also requires the provision of adequate nutrients, aeration and temperature for the growth of microorganisms (VOLESKY, 1990c). However, the fact that many sites where conventional sewage treatment is carried out end up becoming habitats for microorganisms suggests that these limitations do not preclude their application in systems involving the bioaccumulation of heavy metals (DONMEZ & AKSU, 1999).

Microorganisms have the property of adsorbing heavy metals on their cell surface and, among them, fungi and yeasts are more tolerant of toxic metals and can grow in media with high concentrations of these elements. In addition, the small size of microbial cells provides a high ratio between surface area and volume, which implies a large contact surface with the environment and, therefore, a greater possibility of adsorbing or incorporating the metal (KURODA & UEDA, 2003).

Another important characteristic of many yeasts is their ability to thrive in adverse conditions such as extreme pH and high temperatures, tolerating extreme environmental conditions.

The combination of these characteristics, high tolerance to heavy metals and development under adverse conditions, makes yeasts important adsorbents in the removal of toxic metals. It is important to note that microbial biomass is capable of accumulating metals whether it is alive or dead (KAPOOR & VIRARAGHAVAN, 1995).

The abundant fungal biomass produced as a by-product of large-scale industrial processes can be an economically viable source of metal biosorbents. The transformation of waste biomass into a metal biosorbent not only drastically reduces the cost of producing the biosorbent, but also reduces the cost of disposing of waste biomass from industries, such as for fungi used in the production of organic acids (WAIHUNG et al, 1999; KAPOOR & VIRARAGHAVAN, 1995), or for yeasts produced by the Brazilian sugar-alcohol industry, see below.

The National Alcohol Program (PROÀLCOOL) began in Brazil in 1975 and stimulated the development of the sugar-alcohol agro-industry. As a result, many products from this agro-industry began to be produced. Among these products is the large production of *Saccharomyces cerevisiae* yeast, which is used to ferment sugar cane juice. *Saccharomyces cerevisiae* yeast can be obtained from distilleries after it has been used as yeast for alcohol production. For every liter of alcohol produced, about 30 grams of dry weight of yeast are left over. Brazil's annual alcohol production is around 15 billion liters, so yeast waste can be estimated at around 450,000 tons. It is estimated that in 1996/1997 around 25,000 tons of yeast were sold at an average price of approximately R$ 300.00 per ton of yeast (DEL RIO, 2004).

Specifically in the state of Minas Gerais, the cachaça agribusiness plays an important role for thousands of rural properties in the interior of the state. In economic terms, the 8,466 stills in Minas Gerais, according to SEBRAE, generate around 240,000 direct and indirect jobs. According to SEBRAE, the state of Minas Gerais produces around 180 million liters of cachaça every year. It is believed that this figure is underestimated, due to the high rate of clandestinity in the sector (SEBRAE, 2004). Most of Minas Gerais' stills are concentrated in the North, Jequitinhonha and Rio Doce regions, economically deprived areas where any new source of income has a major social impact (SEBRAE, 2004). This data shows how much yeast is produced in the state of Minas Gerais and that the residual biomass could be put to another use.

Industries will only be interested in using biomass if the low cost and efficiency of the process are proven. The low cost will depend on large-scale production, which is why this work has chosen to verify the potential of the biomass produced by cachaça stills. The effectiveness of the methodology will depend on studies proving that yeast biomass removes reasonable amounts of metal, which is why this work was carried out.

2 RELEVANCE AND JUSTIFICATION

The high level of industrial development that has taken place over the last few decades has been one of the main causes of contamination of our water, soil and air, either through negligence in treating it before it is discharged into rivers or through accidents and increasingly frequent carelessness, which lead to the release of many pollutants into all environments.

These pollutants include heavy metals, which cause major problems for human health. Heavy metals are chemical elements with a relatively high atomic weight which, in high concentrations, are very toxic to life. Industrial activities have introduced heavy metals into water in much greater quantities than would be natural, causing pollution. It is enough to remember that heavy metals are part of the discharges of large industries in every country in the world. Among the metals considered most dangerous to human health are mercury, cadmium, chromium and lead. All forms of life are affected by the presence of metals, depending on the dose and chemical form. Many metals are essential for the growth of all types of organisms, from bacteria to humans, but they are required in trace concentrations above which they can damage biological systems.

Cadmium is a heavy metal that is extremely toxic to both humans and the environment. It has been increasingly used in the production of pigments for paints, plastics, batteries and electrolytic deposits. Because of this, the metallic waste must be treated before being disposed of in liquid effluents. Normally, this treatment is carried out using chemical methods, which can be effective, but at a high cost that often makes their use unfeasible. In the search for new alternatives, bioremediation, i.e. the use of biological materials or organisms such as bacteria, fungi, yeasts, algae and plants, has been widely studied.

The detoxification of industrial effluents using biomass will only be possible if the low cost and effectiveness of the process are proven. The low cost of the bioabsorbent material will depend on its large-scale production. Therefore, this study aimed to verify the potential of the yeast biomass produced by cachaça stills, given that this is an economically important activity in the state of Minas Gerais and that there is a large production of yeast by the sugar-alcohol industry throughout the country. In the same way that we used yeasts isolated from the fermentation of cachaça, we also evaluated the ability of yeasts isolated from different regional ecological niches to incorporate cadmium.

The effectiveness of the methodology using yeasts as biosorbent material will depend on studies that prove that these microorganisms remove reasonable amounts of metal, which is why this study measured various parameters in the cells and parameters of metal uptake. The aim of this study was therefore to evaluate the bioremediation potential of yeasts isolated from cachaça fermentation and from regional environments, for use as cadmium-absorbing material in liquid effluents.

3 OBJECTIVES

3.1   General objective

To evaluate the potential of yeasts isolated from the fermentation of cachaça and from regional ecological niches to capture cadmium under laboratory conditions.

3.2   Specific objective

-       Use different yeast strains isolated from the fermentation of cachaça and yeasts of different genera and species, isolated from ecological niches in the region, which may be able to capture a greater amount of cadmium.

-       To verify the influence of exposure to increasing concentrations of cadmium chloride on growth, yeast tolerance and cell trehalose levels.

-       To verify the influence of different cadmium compounds on the incorporation of the metal by the cells.

-       Check the influence of the exposure time of yeast to cadmium on the incorporation of the metal.

-       Determine the influence of cell mass on cadmium uptake.

-       Determine the influence of the yeast life cycle (exponential or stationary phase) on cadmium uptake.

-       Determine the most effective mechanism used by yeast to capture cadmium: passive or active mechanism, using viable and non-viable cells.

-       To compare the strains isolated from the fermentation of cachaça and from regional niches with laboratory yeast strains: *Saccharomyces cerevisiae* W303-WT and *Saccharomyces cerevisiae* S288C in various respects.

4 METHODOLOGY

4.1 Strains studied

This study used 17 different yeast strains, 11 of which were *Saccharomyces cerevisiae*, nine isolated from cachaça-producing distilleries in the state of Minas Gerais. Two strains are laboratory strains: *Saccharomyces cerevisiae* W303-WT and *Saccharomyces cerevisiae* S288C.

The yeasts were obtained from Professor Carlos Augusto Rosa's collection in the Ecology and Biotechnology Laboratory of the Microbiology Department at ICB/UFMG. The entire methodology for collecting, isolating and identifying the yeasts is described in MORAIS et al, 1997 and PATARO et al, 2000. The yeasts were isolated and characterized using conventional methods (YARROW, 1998) and their identities were verified using the taxonomic classification keys of KURTZMAN & FELL 1998.

Table 1: Strains used in this study

Cells	Distillery/Origin	City/Location	Region
Saccharomyces cerevisiae W303-WT	Laboratory strain (*MAT ade2-1-can1-100 trp1-1 his3-11, 15 leu2-3, 112 ura3-1*).	-	
Saccharomyces cerevisiae S288C	Laboratory strain (*MATa his31-leu2 0* *ura3 0 met15 0*).	-	
Saccharomyces cerevisiae 1011	Seleta Boazinha Distillery	Salinas	Jequitinhonha
Saccharomyces cerevisiae 1978	Galo Bravo Distillery	Salinas	Jequitinhonha
Saccharomyces cerevisiae 2041	José Cruz Distillery	Salinas	Jequitinhonha
Saccharomyces cerevisiae 2049	Sebastiao Distillery	Salinas	Jequitinhonha
Saccharomyces cerevisiae 2057	Derci Distillery	Salinas	Jequitinhonha
Saccharomyces cerevisiae 2062	Déia Distillery	Salinas	Jequitinhonha
Saccharomyces cerevisiae 2089	Preciosa Distillery	Salinas	Jequitinhonha
Saccharomyces cerevisiae 2097	Seleta Boazinha Distillery	Salinas	Jequitinhonha
Saccharomyces cerevisiae 2464	Seleta Boazinha Distillery	Salinas	Jequitinhonha
Starmerela meliponinorum UFMG-J26.1	Isolated from the pollen of the bee *Tetragonisca angustula* in a hive	Betim	Rio Doce State Park Rio Doce Park

Pichia guilliermondii DC123.1 Isolated from *Drosophila* spp UFMG Ecological Station Rio Doce State Park *Pichia membranaefaciens* DC 63.3 Isolated from *Drosophila* spp UFMG Ecological Station Rio Doce State Park *Pichiakluyvera* DC 44. 3 Isolated from *Drosophila* spp UFMG Ecological Station Rio Doce State Park

Candida cylindracea DC 44.2 Isolated   from *Drosophila* spp UFMG Ecological Station Rio Doce Park

4.2   Culture Media

All the experiments were carried out in YPG liquid culture medium (2% glucose, 1% yeast extract, 2% peptone). When it was necessary to use plates with solid medium, 2% agar was added to the YPG medium.

4.3   Maintenance of yeast strains in the laboratory

We used three different techniques to store the yeast in the laboratory:

-Storage of the cells in mineral oil in the refrigerator:

A loop containing cells was inoculated into tubes containing solid YPG medium, previously tilted. After 48 hours at 30oC, sterilized mineral oil was added to the surface of the yeasts. The tubes were stored at 4 C.^{0}

-Maintenance in petri dishes:

The yeasts were kept in petri dishes containing solid YPG medium and periodically replanted from the original in an inclined tube filled with mineral oil. The plates were also kept in the refrigerator.

-Maintenance in glycerol medium:

The strains were pre-grown in YPG medium under agitation at 30^0 C, collected in stationary phase by centrifugation and resuspended in medium containing 1% yeast extract, 2% peptone and 25% glycerol. 1 ml of this medium containing the cells was transferred to cryogenic tubes and stored in a freezer at -70 C.^{0}

4.4   Experimental Protocol

The cells were pre-incubated at a temperature of 30^0 C, shaken at 160 rpm (rotation per minute) for a period of approximately 15 hours in YPG liquid medium. The medium was then centrifuged (5 min, 1000 g), the supernatant discarded and the cells transferred to a new YPG liquid medium for a period of 3 hours. We used this 3-hour period to allow the cells to enter the exponential growth phase. The exponential phase of growth was checked by the presence of glucose in the medium. The presence of glucose was checked using glucose indicator strips in the urine. According to experimental requirements, the cells were incubated in YPG medium plus pre-established concentrations of cadmium chloride. Cadmium chloride was autoclaved on its own in a concentrated aqueous solution and then added to the YPG medium. Subsequently, the cells were incubated in YPG medium at 30^0 C, under agitation at 160 rpm, for pre-established periods of time. The incubation time and cadmium chloride concentration varied according to the needs of each experiment. After the established time, the cells were collected by vacuum filtration using 0.45 m pore size nitrocellulose filters, 47 mm in diameter. The cells were washed twice with cold distilled water. The cell mass was removed from the filter with a spatula, transferred to pieces of aluminum foil, frozen in liquid nitrogen and stored in a -20^0 C freezer. This frozen cell mass was then used for the desired dosages.

4.5   Yeast growth curve

For the yeast growth curve, a pre-inoculum of each strain was grown for around 15 hours at 30^0 C, at 160 rpm, in YPG medium. After this period, the optical density of the culture was checked on a spectrophotometer at

660 nm. A sample was inoculated into a new YPG medium and incubated under agitation at 30^0 C. At pre-established times, an aliquot, usually 100 l, was removed and placed in 900 l of water. Growth was followed by measuring the turbidity of the suspension at 660 nm in a spectrophotometer. If necessary, dilutions were made in water to record the turbidity.

4.6 Cadmium tolerance test.

The cells were pre-incubated at a temperature of 30^0 C, at 160 rpm, for a period of approximately 15 hours, in liquid YPG medium. The medium was then centrifuged (5 min, 1000g), the supernatant discarded and the cells transferred to new liquid YPG medium for 3 hours. The cells were then counted using a Neubauer chamber. After counting, 300 l of YPG medium, containing 1×10^6 cells/ml, were placed in the wells of a metal plate. The top of the metal plate, which consisted of a piece of metal rods, was then fitted into each well. After 3 minutes, at room temperature, the top of the plate was removed and placed on YPG solid medium containing increasing concentrations of cadmium chloride. The petri dishes with YPG solids and cadmium chloride remained at room temperature for four days. After this time, the growth of the cells in the plates with cadmium and in the absence of cadmium (control) were compared. The comparison was based on visual inspection to determine which strains had formed colonies on the plates.

4.7 Resistance Induction Tests

To carry out this experiment, the cells were pre-incubated at 30^0 C, 160 rpm, for a period of approximately 15 hours in YPG liquid medium, and transferred by centrifugation to a new YPG liquid medium for a period of 3 hours. The cells were then transferred to YPG medium containing 2.5 ppm cadmium chloride. After 8 hours in these conditions, a further 50 ppm cadmium chloride was added to the medium which already contained 2.5 ppm cadmium chloride. Incubations continued for a further 7 hours at 30^0 C, under agitation. In parallel, control incubations were maintained (absence of cadmium chloride and incubations with 2.5 and 50 ppm cadmium chloride). Subsequently, the turbidity at 660 nm was determined to verify growth and the cells were collected to determine trehalose levels.

4.8 Trehalose dosage

After carrying out the desired experimental protocol, the cells frozen in liquid nitrogen were transferred to conical tubes, kept on ice, and weighed. They were then resuspended in 1 ml of 0.25 M Na_2CO_3, homogenized in a vortex, boiled for 20 minutes and centrifuged for 5 minutes at 1000g. After this step, 200 l of the supernatant was acidified with 1.2 N acetic acid (final concentration 0.30 N) to pH 5.0-5.5 (checked with a pH indicator strip) and buffered with 300 mM sodium acetate containing 30 mM CaCl2 (final concentration 75 mM sodium acetate + 7.5 mM $CaCl_2$). The solution was mixed in a vortex and 100 l of the mixture was used for the determination of trehalose. Trehalose was determined enzymatically with a preparation of trehalase extracted from the fungus *Humicola grisea* var. *thermoidea*. The reaction was carried out for 2 hours at 40°C. After this period, 50 l of this solution was used to measure the glucose formed. The glucose formed was determined using a commercial kit from Bioclin (Glucose GOD - Clin). The reaction took place at 37^0 C for 15 minutes. The colored product formed was determined using a spectrophotometer at 505nm. All dosages were carried out in duplicates. Parallel incubations were carried out for each sample in the presence of denatured trehalase to check for possible glucose contamination of the medium. One unit of trehalase was defined as the amount of enzyme that released 1 mole of glucose/g wet weight.

The accumulated trehalose was quantified using the methodology described by NEVES et al (1994).

4.9 Extraction of trehalase from *Humicola grisea*

Trehalase was extracted from the fungus *Humicola grisea* var. *thermoidea*, grown on oat agar (4% oat flour, 1.8% agar, in tap water) for 20 days at 40^0 C. After this period, the flasks were left for 5 days at a temperature of 4^0 C, a procedure which increases the amount of trehalase to be extracted. 10 ml of distilled water was added to the flasks and the surface of the agar was scraped with a glass rod. The suspension formed was filtered through a funnel containing gauze. The suspension was then centrifuged at 1000g for 40 minutes. 1% Kaolin (hydrated aluminum silicate/Sigma) was added to the supernatant and the mixture was left at room temperature for 20 minutes with sporadic stirring. The solution was centrifuged for 20 minutes at 1000 g, the supernatant was dialyzed every night against H_2O at 4^0 C. The suspension was poured into tubes and frozen. For each batch of enzyme prepared, activity tests were carried out using commercial trehalose (Sigma) as a standard at concentrations of 2, 5 and 10 mM.

4.10 Determination of cadmium incorporated by the cell

We used neutron activation analysis to determine cadmium incorporated into cells. Neutron activation analysis has been recognized as one of the most important analytical tools for determining elemental chemical composition at trace levels. The principle of the technique is to induce artificial radioactivity in a sample by irradiating it with neutrons and then measuring the induced activity by detecting gamma radiation. The physical phenomena on which this analysis is based are the properties of the nucleus, radioactivity and the interaction of radiation with matter via the neutron-gamma reaction (n,). The gamma rays emitted, called decay gamma rays, have energies that are characteristic of each radionuclide. Thus, when detected by gamma spectroscopy, they can be used to identify and quantify the chemical elements present in a sample. Around 70% of natural chemical elements have nuclear properties suitable for neutron activation. Neutron activation analysis is multielement, i.e. the irradiation of the sample and gamma spectroscopy is an inherently multisotopic process, making it possible to determine a large number of elements simultaneously (around 30 elements). The technique has a low incidence of interferences, as various combinations of irradiation time, decay and counting, as well as the selection of different gamma energies for counting after irradiation, can be used. The technique of neutron activation is very selective; elements that are difficult to analyze by conventional analytical techniques are relatively easy to analyze by neutron activation. These include rare earths, precious metals and some toxic elements such as antimony and arsenic. Selectivity is mainly due to the different nuclear properties for elements with similar chemical properties. The sensitivity of the technique is high and depends on many experimental parameters that can be adjusted, as well as nuclear parameters and the neutron flux in the reactor. Due to the high sensitivity, small amounts of sample are required, in some cases just a few milligrams. This is a great advantage when analyzing small quantities of samples or precious samples.

The technique is capable of analyzing solid, liquid and gaseous matrices: the nuclear reaction (n,) is independent of the physical state of the matrix. The method is non-destructive and the sample is not visually or chemically altered.

Neutron activation analysis, when properly performed, is one of the most precise and accurate analytical methods. This makes it an important analytical tool for certification, calibration and intercomparison of samples in trace and ultra-trace levels.

Over the years, this technique has made remarkable progress due to computer advances, automatic sample handling, the development of intrinsic germanium detectors and appropriate electronic instruments (IAEA-TECDOC 435, 1987; IAEA-TECDOC 564, 1990; EHMANN & VANCE, 1991; KRUGER, 1971; DE SOETE et al, 1972).

Neutron activation is carried out in research reactors. In Brazil, only two research centers use neutron activation, the Nuclear Technology Development Center (CDTN) in Belo Horizonte and the Nuclear and Energy Research Institute (IPEN) in Sao Paulo, both belonging to the National Nuclear Energy Commission (CNEN).

The Nuclear Technology Development Center has the TRIGA IPR-1 reactor, equipped with a rotating table and an irradiation device especially suitable for this analytical technique. During irradiation, the table rotates around the reactor core at a constant speed to ensure a uniform neutron flux in the samples. Currently, the reactor's power is 100 kW and at this power the neutron flux reaches a value of around $6.6 \ 10 \ \mathrm{n/cm}^{11 \ -2} \ \mathrm{s}^{-1}$. The half-life of[115] Cd is 2.2 days (MENEZES et al, 2003).

5 RESULTS

Our first approach was to see what influence the presence of cadmium had on the growth curve of the yeast *Saccharomyces cerevisiae* W303-WT, a laboratory strain.

The cells of the *Saccharomyces cerevisiae* W303-WT strain in the control condition (no cadmium) showed the expected growth profile: an increase in the number of cells as time progressed, the progressive increase in the concentration of cadmium used (2 ppm to 50 ppm) reduced the growth of the yeasts (Figure 1). After 2 hours the control cells grew, remaining in exponential phase for a period of 20 hours. Cells exposed to 2 ppm cadmium chloride had their growth reduced compared to the control cells, but compared to the other concentrations (5 ppm to 50 ppm), cells exposed to 2 ppm cadmium chloride grew faster. In the presence of 50 ppm cadmium chloride, growth was slow, and after 22 hours the cells had grown by about 25% of that obtained by cells incubated in the absence of cadmium chloride (control).

After verifying that the *Saccharomyces cerevisiae* W303-WT strain responds to the presence of cadmium chloride with a pronounced effect on growth, it was decided to see if the presence of cadmium had an influence on the accumulation of trehalose (Figure 2).

Figure 2 shows that the levels of trehalose in the cells increased in the presence of cadmium chloride, when compared to the cells in control conditions (absence of cadmium chloride). It was observed (Figure 2) that as the concentration of cadmium increased, the levels of trehalose increased.

The cells in the control condition were collected in the exponential phase of growth and therefore did not have high trehalose levels. It can be seen, however, that the presence of cadmium chloride caused an increase in trehalose levels when increasing concentrations of this chemical compound were added.

When the cells were exposed to concentrations of cadmium chloride ranging from 50 ppm to 850 ppm, the levels of trehalose increased progressively as the concentration of cadmium chloride increased. At a concentration of 1000 ppm, trehalose levels were lower than those obtained at concentrations of 850 ppm, 750 ppm and 500 ppm. Such

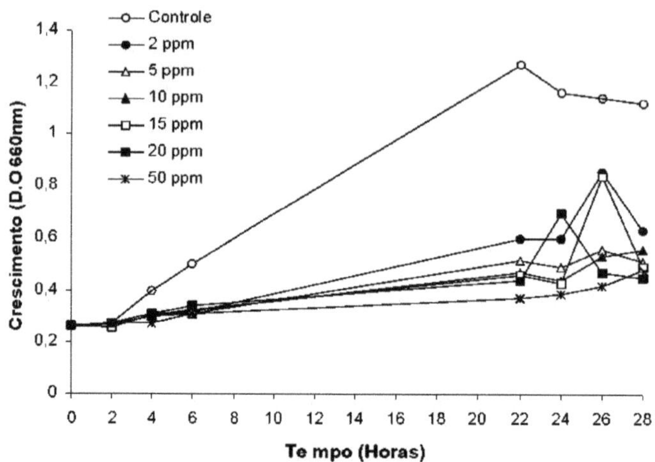

Figura 1 - Influence of different concentrations of cadmium chloride on the growth of the yeast *Saccharomyces cerevisiae* W303-WT (laboratory strain).

The cells of the *Saccharomyces cerevisiae* W303-WT yeast strain were incubated at 30^0 C, under agitation (160 rpm) for around 15 hours in YPG medium. They were then transferred by centrifugation to a new YPG medium and left for 3 hours under the same conditions as above. After this period, the medium was divided and the appropriate cadmium concentration was added to each part. The control condition was carried out in the absence of cadmium chloride. The cells remained incubated in the presence of cadmium chloride for the specified times, when a sample was taken for turbidity analysis. Determinations were made using a spectrophotometer at 660nm.

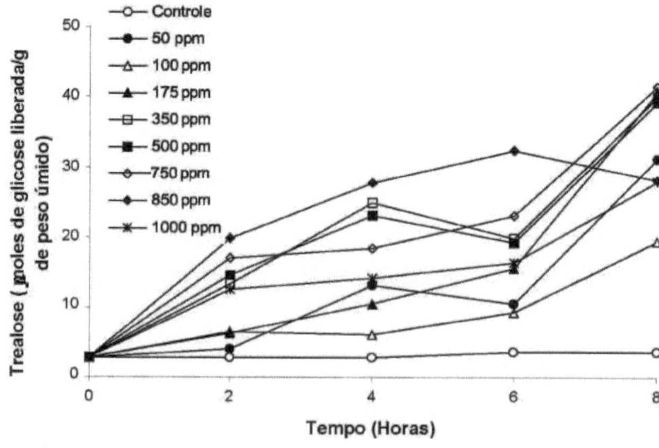

Figura 2 - Influence of increasing concentrations of cadmium chloride on trehalose levels.

The cells of the yeast strain *Saccharomyces cerevisiae* W303-WT were incubated at 30^0 C, under agitation

26

(160 rpm) for about 15 hours, in YPG medium. Subsequently, the cells were collected by centrifugation and transferred to a new YPG medium for 3 hours under the same conditions specified above. After this period, the medium was divided and the appropriate concentrations of cadmium chloride were added to each part. The control condition was carried out in the absence of cadmium chloride. The cells remained in these conditions for a period of 8 hours. At the specified times, samples were filtered, frozen in liquid nitrogen and stored in a freezer. Subsequently, the cell mass was extracted to measure trehalose, according to a previously described methodology. The trehalose values were expressed in moles of glucose released per gram of wet weight.

The result showed that the higher the concentrations of cadmium chloride used, the higher the levels of trehalose accumulated. Exposure of the cells to a concentration of 1000 ppm cadmium chloride was very toxic to the cells, and it was not possible to synthesize trehalose.

The next step was to check which concentration of cadmium allowed the cells to grow and reproduce. Table 2 shows that all the yeasts used were able to form colonies at concentrations of up to 15 ppm cadmium chloride, with the exception of the laboratory strain *Saccharomyces cerevisiae* W303-WT.

The *Pichia kluyvera* and *Pichia guilliermondii* strains did not grow at a concentration of 50 ppm cadmium chloride.

The strains that grew at concentrations of up to 50 ppm cadmium chloride were: *Starmerela meliponinorum, Candida cylindracea, Pichia membranaefaciens*, and all the *Saccharomyces cerevisiae* yeast strains isolated from cachaça fermentation: *Saccharomyces cerevisiae* 1011, *Saccharomyces cerevisiae* 1978, *Saccharomyces cerevisiae* 2041, *Saccharomyces cerevisiae* 2049, *Saccharomyces cerevisiae* 2057, *Saccharomyces cerevisiae* 2062, *Saccharomyces cerevisiae* 2089, *Saccharomyces cerevisiae* 2097, *Saccharomyces cerevisiae* 2464 (Table 2).

The only strain that grew and reproduced in the presence of up to 100 ppm cadmium chloride was the yeast *Starmerela meliponinorum*.

None of the strains tested grew at concentrations of 500 ppm, 1000 ppm and 2000 ppm cadmium chloride (results not shown).

After finding that cadmium chloride caused a decrease in cell culture growth (Figure 1), stimulated the synthesis of trehalose (Figure 2) and interfered with the division and reproduction of the various yeast strains (Table 2), experiments were carried out to see if the chemical form in which cadmium was presented to the cells had an influence on trehalose levels (Figure 3) and its incorporation (Figure 4). To this end, 300 ppm of each of the following chemical compounds was added to the incubation medium containing *Saccharomyces cerevisiae* W303-WT cells: cadmium chloride, cadmium oxide, cadmium acetate, cadmium nitrate and cadmium sulphate.

It can be seen in Figure 3 that trehalose levels increased in cells exposed to 300 ppm of the five types of cadmium salts, compared to the data obtained from control cells (no cadmium salts).

Table 2 - Tolerance test of yeasts to cadmium

Strains / ppm $CdCl_2$	0	5	10	15	20	50	100	250
Saccharomyces cerevisiae W303-WT	C	C	C	N	N	N	N	N
Saccharomyces cerevisiae 1011	C	C	C	C	C	C	N	N
Saccharomyces cerevisiae 1978	C	C	C	C	C	C	N	N
Saccharomyces cerevisiae 2041	C	C	C	C	C	C	N	N
Saccharomyces cerevisiae 2049	C	C	C	C	C	C	N	N
Saccharomyces cerevisiae 2057	C	C	C	C	C	C	N	N
Saccharomyces cerevisiae 2062	C	C	C	C	C	C	N	N
Saccharomyces cerevisiae 2089	C	C	C	C	C	C	N	N
Saccharomyces cerevisiae 2097	C	C	C	C	C	C	N	N
Saccharomyces cerevisiae 2464	C	C	C	C	C	C	N	N
Starmerela meliponinorum	C	C	C	C	C	C	C	N
Candida cylindracea	C	C	C	C	C	C	N	N
Candida sorboxylosa	C	C	C	C	C	C	N	N
Pichia guilliermondii	C	C	C	C	C	N	N	N
Pichia kluyvera	C	C	C	C	C	N	N	N
Pichia membranaefaciens	C	C	C	C	C	C	N	N

Legend: C= growth, N= no growth.

The yeast cells of the different strains were incubated at 30^0 C, under agitation (160 rpm), for about 15 hours in YPG medium. After this period, they were centrifuged and transferred to a new medium (YPG) under the same conditions specified above for 3 hours. After this period, 1 x 10^6 cells/ml were placed in the well plate and plated on solid YPG medium containing increasing concentrations of cadmium chloride (5 ppm á 250 ppm). The plates were left at room temperature for four days. The growth of the colonies on the plates was analyzed by visual inspection, comparing the growth of the colonies on the control plate with the growth of the colonies exposed to different concentrations of cadmium chloride.

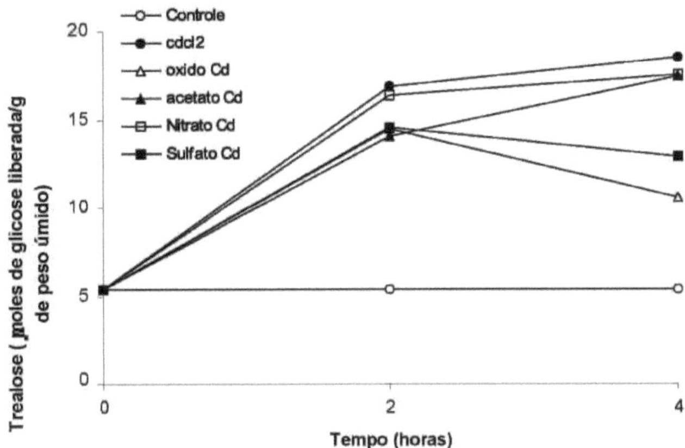

Figura 3 - Determination of the amount of trehalose accumulated in the yeast *Saccharomyces cerevisiae* **W303-WT after exposure to different cadmium compounds**

The cells of the yeast strain *Saccharomyces cerevisiae* W303-WT were incubated at 30^0 C, under agitation (160 rpm) for about 15 hours in YPG medium. After this period, the medium was divided and 300 ppm of the following chemical reagents were added to each part: cadmium chloride, cadmium oxide, cadmium acetate, cadmium nitrate and cadmium sulphate. The control condition was carried out in the absence of cadmium. The cells remained in these conditions for a period of 4 hours. At the specified times, samples were filtered, frozen in liquid nitrogen and stored in a freezer. Subsequently, the cell mass was extracted to measure trehalose, according to a previously described methodology. The trehalose values were expressed in moles of glucose released per gram of wet weight.

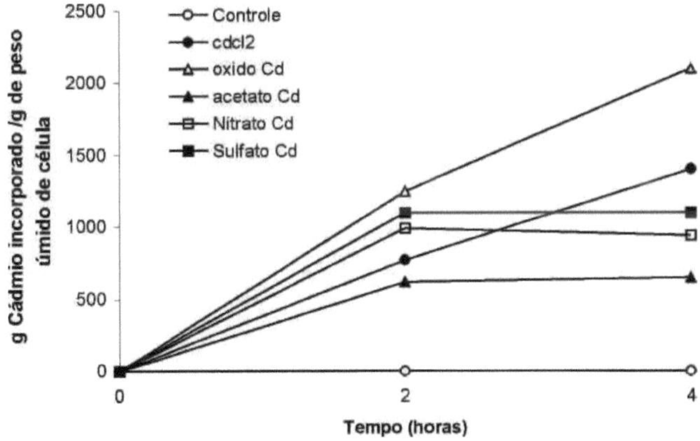

Figura 4 - **Determining the amount of cadmium incorporated into** *Saccharomyces cerevisiae* **strain W303-WT**

The cells of the yeast strain *Saccharomyces cerevisiae* W303-WT were incubated at 30^0 C, under agitation

(160 rpm) for about 15 hours in YPG medium. The cells were then collected by centrifugation and transferred to a new YPG medium for 3 hours under the same conditions as above. After this period, the medium was divided and 300 ppm of the following chemical reagents were added to each part: cadmium chloride, cadmium oxide, cadmium acetate, cadmium nitrate and cadmium sulphate. The control condition was carried out in the absence of cadmium. The cells remained in these conditions for a period of 4 hours. At the specified times, samples were filtered, washed three times with distilled water, frozen in liquid nitrogen and stored in a freezer. The cell mass was then weighed and sent to the Triga reactor for irradiation and subsequent determination of the amount of cadmium incorporated by the cells. The values were expressed as grams of cadmium incorporated per gram of wet cell weight.

Next, the amounts of cadmium incorporated by the cells were determined when the metal was presented in different chemical formulas (Figure 4). All five different cadmium salts were incorporated by the *Saccharomyces cerevisiae* W303-WT yeast cell strain.

Cadmium chloride and cadmium oxide provided relatively greater incorporation of the metal over the time studied. Cadmium chloride also had an advantage in inducing trehalose synthesis over the other cadmium salts studied. However, it was decided to use cadmium chloride in all the experiments, which will be described below.

To check whether it was possible to induce resistance to cadmium chloride, cells from the *Saccharomyces cerevisiae* W303-WT strain were used. The cells were incubated in the absence of cadmium chloride and in the presence of 2.5 ppm and 50 ppm cadmium chloride for 15 hours (Figure 5A and Figure 5B).

In the 8-hour period, the cells in the control condition grew without any problems (Figure 5A). In the presence of 2.5 ppm cadmium chloride, growth was inhibited by around 27%. In the presence of 50 ppm cadmium chloride, growth was inhibited by around 55% compared to the growth of cells in the control condition. After this 8-hour period, the incubation carried out in the presence of 2.5 ppm cadmium chloride was divided in half. Part of it remained in the same condition (2.5 ppm) and 50 ppm cadmium chloride was added to the other part. The cells that received a further 50 ppm of cadmium chloride had reduced growth compared to those that remained in 2.5 ppm of cadmium chloride, but this reduction in growth did not reach the levels determined for the growth of the cells that remained in 50 ppm of cadmium chloride throughout the experiment.

This result showed that pre-treatment with low concentrations of cadmium chloride gave the cells an advantage when they were exposed to higher concentrations of this metal. Compared to the control, it was observed that the presence of 2.5 ppm cadmium chloride caused trehalose levels to be higher.

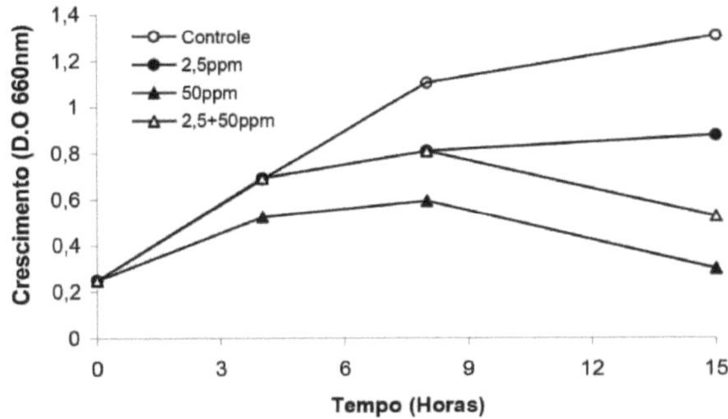

Figure 5 A- Pre-adaptation of yeast cells in the presence of cadmium chloride.

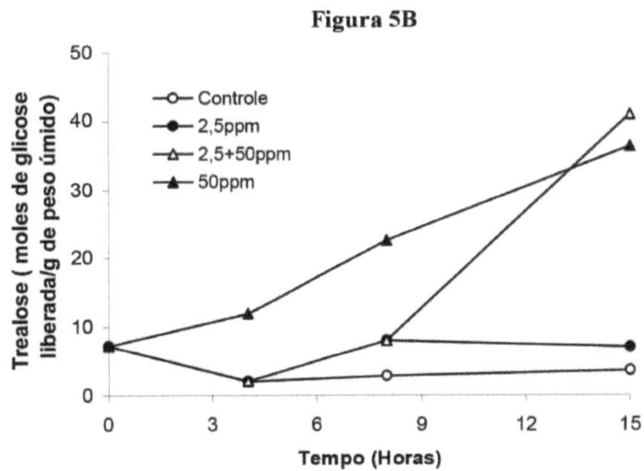

Figure 5B - Levels of trehalose accumulated by cells incubated in the presence of cadmium.

The yeast strain Saccharomyces cerevisiae W303-WT was incubated at 30^0 C, under agitation (160 rpm) for about 15 hours, in YPG medium. It was then transferred by centrifugation to a new YPG medium and left for 3 hours under the same conditions as specified above. After this time, the medium was divided into 3 parts: one part remained in YPG (control), one part was added 2.5 ppm cadmium chloride and the other part was added 50 ppm cadmium chloride. After 8 hours in these conditions, the incubation with 2.5 ppm cadmium chloride was subdivided. Part remained in the same condition, i.e. with 2.5 ppm cadmium chloride, and the other part was given another 50 ppm cadmium chloride. The cells were incubated under the conditions described above and samples were taken for turbidity analysis at the specified times. Determinations were carried out using a spectrophotometer at 660nm (part A) and for determining accumulated trehalose levels (part B). The trehalose values were expressed in moles of glucose released per gram of wet weight.

In this same experiment, in which the growth of cells adapted to low concentrations of cadmium chloride was verified, samples were taken to determine the levels of trehalose in these same cells (Figure 5A and Figure

5B). As expected, the control cells (absence of cadmium chloride) did not synthesize trehalose, as they were not subjected to any stress during this time, were in the presence of glucose and were in the exponential phase of growth. The increase in trehalose levels was low when the *Saccharomyces cerevisiae* W303 -WT cells were incubated in the presence of low concentrations of cadmium chloride (2.5 ppm), however, cells exposed to a higher concentration of cadmium chloride (50 ppm), when compared to those incubated directly with 2.5 ppm cadmium chloride, had a marked increase in trehalose levels.

The cells that were exposed for 8 hours to 2.5 ppm cadmium chloride and then added another 50 ppm of the metal showed a marked increase in trehalose levels compared to the cells that remained in 2.5 ppm cadmium chloride for 15 hours, but compared to the cells that were incubated with 50 ppm cadmium chloride for the same period of time, the increase in trehalose levels was irrelevant.

The results obtained using *Saccharomyces cerevisiae* W303-WT cells (laboratory strain) were clear: the presence of cadmium at a concentration of 2.5 ppm reduced growth by around 40% in 15 hours (Figure 5A) and the presence of cadmium at any of the concentrations used increased the synthesis of trehalose (Figure 2). We were interested to find out how other yeasts isolated from the fermentation of cachaça and from our region would react to cadmium, because it is well known that microorganisms used in fermentation processes are subject to great fluctuations in chemical, physical and biological parameters and are very resistant to all stresses. For example, during the fermentation of sugarcane juice for the production of cachaça, the yeasts come into contact with the large amount of sugar present in the must, and these high levels of sucrose promote a strong osmotic stress in these cells. Other examples of stresses that occur during the fermentation of cachaça are thermal stress, when the temperature in the fermentation vats can reach 41^0 C during the summer, and alcoholic stress, when, as the fermentation develops, the alcohol content increases. Despite being subject to these stresses, yeasts do very well in fermentation because they are adapted to this task. Based on this observation, the idea was as follows: could the use of yeasts isolated from the fermentation of cachaça, or isolated directly from the environment, be more resistant to the presence of cadmium? To this end, we carried out the experiment shown in Figure 6, where we checked the influence of 500 ppm cadmium chloride for 3 hours on the growth of different yeast genera and species.

Figure 6 showed that 500 ppm cadmium chloride reduced growth in all the strains, but did not inhibit it completely. Some strains such as *Saccharomyces cerevisiae* 1011 and *Saccharomyces cerevisiae* 2041 were more sensitive to this concentration of cadmium chloride, while others of the same genus and species were more resistant (e.g. *Saccharomyces cerevisiae* 1978). The *Pichia guilliermondii* strain showed slight growth at 500 ppm cadmium chloride, but the *Pichia kluyvera* strain had its growth affected by this concentration of cadmium chloride. The *Candida cylindracea* and *Candida sorboxylosa* strains showed higher growth than the *Pichia* strains, remaining closer to the growth of the *Saccharomyces* strains.

The ability to grow at this concentration of cadmium chloride is not related to the genus or species used, and appears to be an accidental phenomenon for the strain in question.

The experiment shown in Figure 7 was carried out to check whether yeasts isolated from the environment also respond to the presence of cadmium by increasing trehalose levels, just like the laboratory strain (Figure 2).

In the *Saccharomyces cerevisiae* strains, there was practically no trehalose synthesis in the cells incubated in the presence of 500 ppm cadmium chloride compared to the control cells (absence of cadmium chloride), but the *Starmerela meliponinorum, Pichia guilliermondii, Pichia membranaefaciens, Pichia kluyvera, Candida sorboxylosa and Candida cylindracea* showed high levels of trehalose compared to the control cells and the other *Saccharomyces* strains (Figure 7).

We conclude from this result that the high concentration of cadmium chloride (500 ppm) made it difficult for some strains to accumulate high levels of trehalose. However, it was possible to experimentally increase trehalose levels in some of the strains using 500 ppm cadmium chloride during the 3 hours of exposure to this chemical compound.

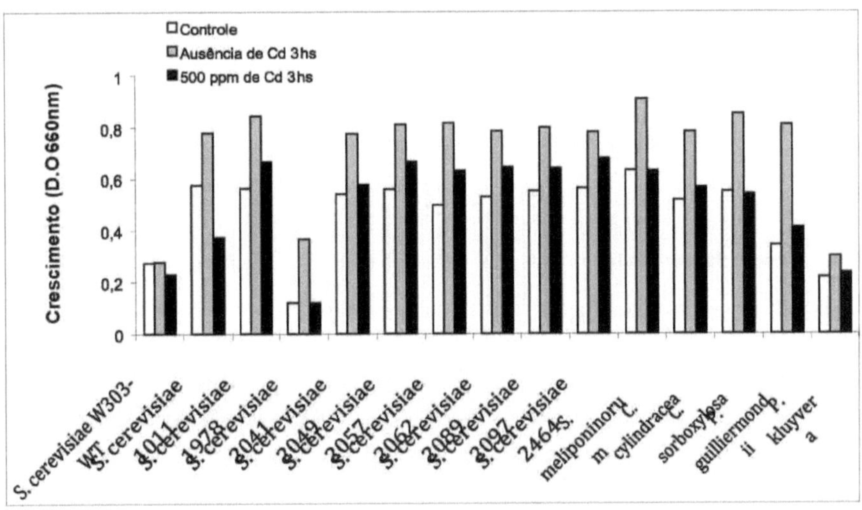

Figure 6 - Influence of 500 ppm cadmium on the growth of different yeasts.

The cells from the different yeasts were incubated at 30^0 C, under agitation (160 rpm) for about 15 hours in YPG medium. After this period, the cells were transferred by centrifugation to a new YPG medium and left for 3 hours under the above conditions. At the end of this period, samples were taken to determine the O.D. (control), and the media were divided. Part of the incubation received 500 ppm cadmium chloride and part was left without the addition of cadmium chloride (Absence of cadmium chloride 3 hours). The cells remained in these conditions for 3 hours when samples were taken for turbidity analysis. Determinations were made using a spectrophotometer at 660nm.

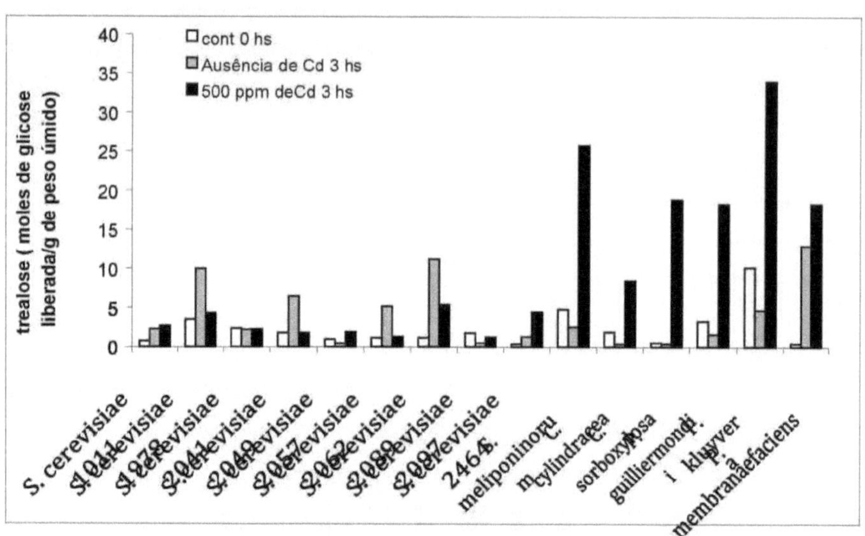

Figure 7 - Influence of 500ppm cadmium chloride on the trehalose levels of different yeast strains.

The cells of the different yeasts were incubated at 30^0 C, under agitation (160 rpm) for about 15 hours in YPG medium. After this period, the cells were transferred by centrifugation to a new YPG medium and left for 3 hours under the above conditions. At the end of this period, samples were taken to measure trehalose (control), and the media were divided. Part of the incubation received 500 ppm cadmium chloride, and part followed without the addition of cadmium chloride (Absence of cadmium chloride 3 hours). The cells remained in these conditions for 3 hours when samples were collected by filtration, frozen in liquid nitrogen and stored in a freezer. Subsequently, the cell mass was extracted to measure trehalose, according to a previously described methodology. The trehalose values were expressed in moles of glucose released per gram of wet weight.

After verifying that yeasts isolated from cachaça fermentation and from the environment are generally more resistant to cadmium, they are able to grow and divide at higher concentrations of this metal, When compared to the laboratory yeast strain *Saccharomyces cerevisiae* W303-WT (Table 2), we carried out a series of experiments to find out about the ability of cells from different yeast strains to capture cadmium from the extracellular medium and accumulate the metal inside the cell.

Firstly, it was checked how much cadmium the *Saccharomyces cerevisiae* W303-WT cells can take up from the extracellular medium and whether this incorporation was dependent on time and the concentration of cadmium present in the medium. The results are shown in Figure 8. The cells in control conditions did not accumulate cadmium, as they were not exposed to this metal. When they were exposed to concentrations ranging from 100 ppm to 1500 ppm of cadmium, we found that the increase in cadmium concentration, in relation to the increase in time, led to greater cadmium incorporation (Figure 8). The conclusion is that the increase in cadmium uptake was dependent on the concentration of the metal and the time that the yeast cells remained in contact with this metal.

In order to check whether the proportion of cell mass is an important factor for the cells to incorporate cadmium, experiments were carried out with cells from the *Saccharomyces cerevisiae* S288C yeast strain (laboratory strain) using increasing cell masses, which can be seen in Table 3. The different quantities of cells were incubated with the same concentration of cadmium (300 ppm) in the same volume (50 ml) for a period of 24 hours. It was observed that, despite the great difference between the masses of cells, the levels of cadmium incorporated were similar, with little variation. Apparently, the quantity of cells does not seem to be a decisive factor in the incorporation of the metal.

After carrying out experiments evaluating the incorporation of cadmium into laboratory yeast strains (Figure 8 and Table 3), experiments were carried out to verify the incorporation of cadmium into various yeast strains. We checked whether yeast cells isolated from cachaça fermentation or from the environment (Figure 9) were able to capture more cadmium than the laboratory yeast strain (Figure 9).

The cells that remained in contact with cadmium for 3 hours incorporated less cadmium than the cells that remained in contact with the metal for 24 hours.

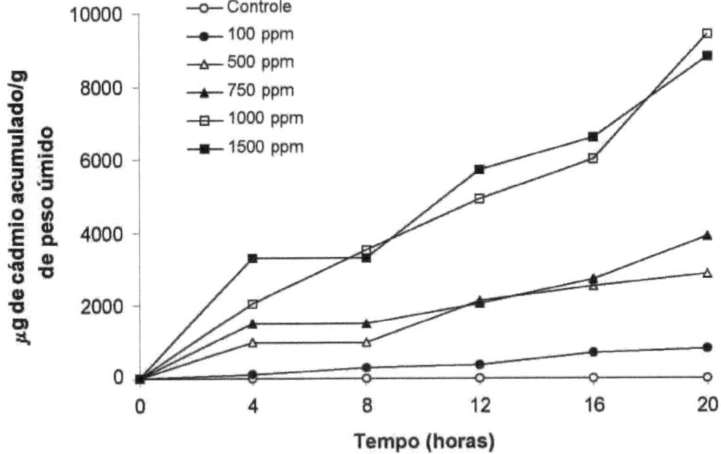

Figura 8 - **Amount of cadmium incorporated by** *Saccharomyces cerevisiae* **W303-WT cells (laboratory strain)**

The yeast strain *Saccharomyces cerevisiae* W303-WT was incubated at 30^0 C, under agitation (160 rpm) for about 15 hours in YPG medium. The cells were then collected by centrifugation and transferred to a new YPG medium for 3 hours under the same conditions as above. After this period, the medium was divided and the specified concentrations of cadmium chloride were added to each part. The control condition was carried out in the absence of cadmium. The cells remained in these conditions for the following times: 4, 8, 12, 16 and 20 hours when the samples were filtered, washed three times with distilled water, frozen in liquid nitrogen and stored in a freezer. The cell mass was then weighed and sent to the Triga reactor for irradiation and subsequent determination of the amount of cadmium incorporated by the cells. The values were expressed in grams of cadmium incorporated per gram of wet cell weight.

Table 3 - Influence of increasing cell masses on cadmium incorporation.

Time (24 hours)	Control	0.77 grams of cells	1.5 grams of cells	2.8 grams of cells	5 grams of cells
Accumulated Cd	ND	200	180	200	200

Legend: ND = no cadmium detected.

The *Saccharomyces cerevisiae* S288C yeast strain was incubated at 30^0 C, under agitation (160 rpm) for about 15 hours, in YPG medium. They were then transferred to 50 ml of YPG medium and left for 3 hours. After this period, a concentration of 300 ppm cadmium was added. The cells remained in these conditions for 24 hours when the samples were filtered, washed three times with distilled water, frozen in liquid nitrogen and stored in a freezer. The cell mass was then weighed and sent to the Triga reactor for irradiation and subsequent determination of the amount of cadmium incorporated by the cells. The values were expressed in grams of cadmium incorporated per gram of wet cell weight.

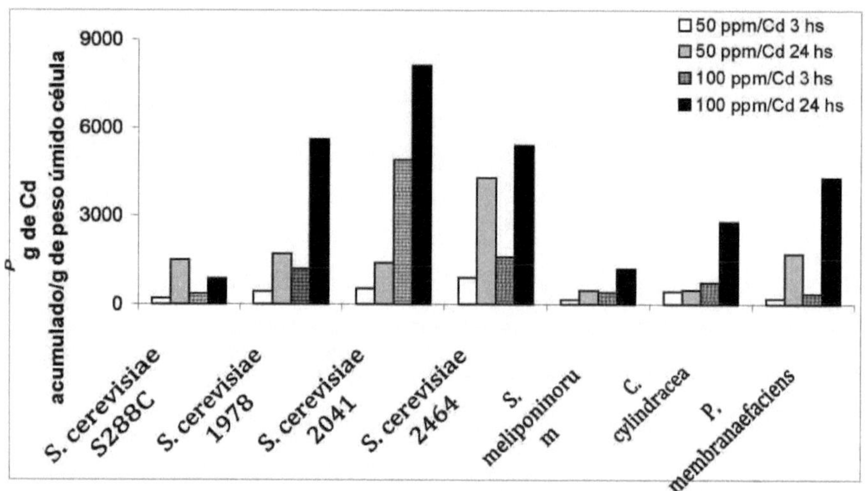

Figura 9   - Influence of cadmium incubation time on metal incorporation.

Cells from different yeast strains were pre-incubated for about 15 hours in liquid YPG medium, under agitation (160 rpm), at 30^0 C. They were transferred by centrifugation to a new YPG medium and left in these conditions for 3 hours. After this period, the incubations were divided into three parts: one part was kept under the same conditions (control), one part was added 50 ppm and the other 100 ppm of cadmium chloride. After 3 hours and 24 hours, samples were filtered, washed three times with distilled water, frozen in liquid nitrogen and stored in the freezer. The cell mass was then weighed and sent to the Triga reactor for irradiation and subsequent determination of the amount of cadmium incorporated by the cells. The values were expressed in grams of cadmium incorporated per gram of wet cell weight. No cadmium was detected in the control condition (no cadmium).

In this experiment, the yeast strains that showed the highest cadmium incorporation were: *Saccharomyces cerevisiae* 2041, then *Saccharomyces cerevisiae* 1978 and *Saccharomyces cerevisiae* 2064. When all the strains are compared, it can be seen that the strains with the lowest incorporation under the conditions used were *Saccharomyces cerevisiae* S288C (laboratory strain) and *Starmerela meliponinorum* (Figure 9).

In Figure 9, seven different yeast strains were incubated with concentrations of 50 ppm and 100 ppm cadmium. It was observed that the concentration of cadmium present in the extracellular medium had an influence on the ability of the different yeast strains to incorporate the metal. It was also possible to verify that this uptake was time-dependent, the longer the time, the greater the incorporation.

The results presented in Figure 9 showed variations in cadmium incorporation that could perhaps be dependent on the growth phase. To observe the phenomenon in more detail, we carried out the experiment shown in Figure 10, where the yeast cells were collected in the stationary phase and only then came into contact with cadmium. We can see that when the cells were inoculated in the stationary phase in the presence of cadmium, the incorporation of cadmium by most yeasts was lower than when the cells were grown in the presence of this metal (Figure 9). With the exception of *Saccharomyces cerevisiae* 2041, which incorporated a considerable amount of cadmium compared to the other strains under these conditions.

Cadmium incorporation seems to depend on the stage of growth the cells are in (Figure 9 and Figure 10).

In order to compare the type of incorporation mechanism being used by the yeast cells (passive or active), experiments were carried out with dead cells after autoclaving. The non-viable biomass was incubated in a medium containing 50 ppm cadmium for 48 hours (Table 4).

As can be seen in Table 4, the cells of the different strains of non-viable yeast were incubated in the presence of 50 ppm cadmium, with the aim of analyzing the type of incorporation processed by these strains: whether it was passive incorporation (a method independent of the microorganism's metabolic cycle, which characterizes biosorption - VOLESKY 1990a) or active incorporation (depending on the cell's metabolism, transport of the metal into the cytoplasm and organelles, which characterizes bioaccumulation - VOLESKY 1990a). Compared to the results in Figure 10 (cell in stationary phase)

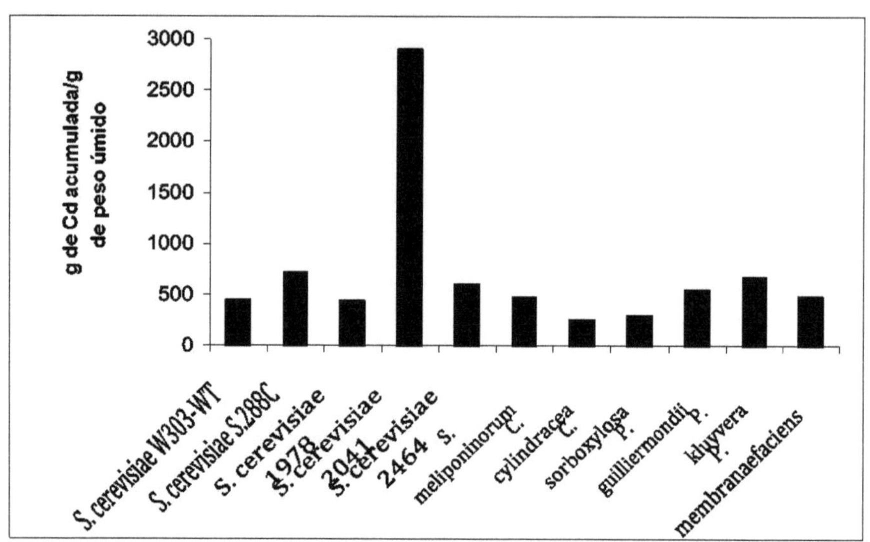

Figura 10 - **Cadmium incorporation in different yeast strains collected in stationary phase.**

The cells of the various yeast strains were incubated for about 48 hours in YPG medium, under agitation (160 rpm) at 30^0 C. After this period, the cultures were divided: part remained in the same condition (control) and 50 ppm cadmium chloride was added to the other part. The cells remained in these conditions for another 12 hours and were then collected by filtration, washed three times with distilled water, frozen in liquid nitrogen and stored in the freezer. The cell mass was then weighed and sent to the Triga reactor for irradiation and subsequent determination of the amount of cadmium incorporated by the yeast cells. The values were expressed in grams of cadmium incorporated per gram of wet cell weight. No cadmium was detected in the control condition (no cadmium).

Table 4 - Incorporation of cadmium by non-viable biomass in different yeast strains

Cells	Control	50 ppm $CdCl_2$
Saccharomyces cerevisiae W303-WT	ND	3400
Saccharomyces cerevisiae 1011	ND	6900
Saccharomyces cerevisiae 1978	ND	5600
Saccharomyces cerevisiae 2041	ND	4000
Saccharomyces cerevisiae 2049	ND	4600
Saccharomyces cerevisiae 2057	ND	5500
Saccharomyces cerevisiae 2062	ND	5000
Saccharomyces cerevisiae 2089	ND	6100
Saccharomyces cerevisiae 2097	ND	4700
Saccharomyces cerevisiae 2464	ND	5500

Starmerela meliponinorum	ND	6200
Candida cylindracea	ND	3800
Candida sorboxylosa	ND	4200
Pichia guilliermondii	ND	5100
Pichia kluyvera	ND	2400
Pichia membranaefaciens	ND	3000

Legend: ND = no cadmium detected

The different yeast strains were pre-incubated in liquid YPG medium for about 15 hours, under 160 rpm shaking at 30⁰ C. They were then transferred to a new YPG medium and left in these conditions for 3 hours. They were then autoclaved, centrifuged and incubated in new YPG medium with 50 ppm cadmium. The cells were incubated for 48 hours, collected by filtration, washed three times with distilled water, frozen in liquid nitrogen and stored in a freezer. The cell mass was then weighed and sent to the Triga reactor for irradiation and subsequent determination of the amount of cadmium incorporated by the cells. The values were expressed in grams of cadmium incorporated per gram of wet cell weight.

and those in Figure 9 (cells in exponential and stationary phase), we can see that the incorporation of cadmium by the dead cells was greater than that of the live cells in the previous experiments, when a concentration of 50 ppm of cadmium was used.

To make it easier to understand the results and discussions, some of the results presented below will be compared.

Table 5 shows the incorporation of cadmium by various strains collected in the exponential growth phase and incubated with three different concentrations of cadmium (50 ppm, 100 ppm and 500 ppm). It was observed (Table 5) that the increase in cadmium concentration led to an increase in metal incorporation for all the strains when they were placed in the presence of increasing cadmium concentrations (50 ppm and 100 ppm). When the cells were incubated for six hours in the presence of 500 ppm cadmium, the same phenomenon was not repeated for the yeast strains *Saccharomyces cerevisiae* 1978 and *Saccharomyces cerevisiae* 2464, showing low cadmium incorporation compared to the data obtained after exposing the cells to 50 ppm and 100 ppm cadmium.

Comparing the incubation time during which the cells remained in contact with cadmium (Table 6), it can be seen that the longer the cells remained in contact with cadmium, the greater the incorporation of the metal in all the strains. The yeast strains *Saccharomyces cerevisiae* incorporated the most cadmium in this situation, followed by the yeast strain *Pichia membranaefaciens* and lastly the laboratory strain *Saccharomyces cerevisiae* S288C.

Table 7 shows that the longer the cells were exposed to cadmium, the more the cells incorporated the metal. With the exception of the *Saccharomyces cerevisiae* 2041 strain, which incorporated more metal in 12 hours of incubation than when incubated for 24 hours. The non-viable cells do indeed seem to have a greater capacity to incorporate the metal, as the exposure time to cadmium under these conditions was 12 hours, but all the strains incorporated much more cadmium in this situation than when they were exposed alive for a period of 24 hours. Table 7 shows the influence of 50 ppm cadmium on different yeast strains when the cells were collected in three different situations. The conclusion is that when the different yeast strains were exposed to the same concentration of cadmium (50 ppm), but in different situations, the condition in which the cells incorporated the most cadmium was when they were exposed to the metal after being dead. Table 7 shows that the situation in which the cells incorporated the least cadmium was when they were exposed to this metal for a period of 3 hours.

The ability of dead cells to take up heavy metals has been extensively studied. It is already known that the ability of dead cells to incorporate metals is greater than that of living cells (KAPOOR & VIRARAGHAVAN, 1995). The use of dead biomass is more viable for industries, as it offers greater advantages over the use of live biomass. Live cells are more sensitive when exposed to metal, and several factors interfere with their viability: the toxicity of the metal, the pH of the medium and the temperature are all factors that interfere with the efficiency of live cells in capturing more metal than dead ones. The biosorption of metals with living cells, however, is more complicated.

Table 8 shows that the incubation of non-viable cells in the presence of cadmium seems to be a favorable situation for the cells to incorporate the metal, because if we compare it with a longer time of exposure to the metal (24 hours) and a higher concentration of cadmium (100 ppm), the biomass made up of dead cells incorporates more cadmium than in another situation.

Table 5 - Influence of exposure to increasing cadmium concentrations on metal incorporation by different strains.

Cells	50 ppm 3 hrs	100 ppm 3hs	500 ppm 6 hs
Saccharomyces cerevisiae S288C	210	350	ND
Saccharomyces cerevisiae 1978	440	1200	250
Saccharomyces cerevisiae 2041	530	4900	12000
Saccharomyces cerevisiae 2464	910	1600	110
Starmerela meliponinorum	160	410	6000
Pichia membranaefaciens	190	350	ND
Candida cylindracea	440	730	4600

ND = Not Determined cadmium.

The values were expressed in grams of cadmium incorporated per gram of wet weight of cells.

The data and experimental conditions for 50 ppm 3 hours and 100 ppm 3 hours were taken from Figure 9 and those for 500 ppm from Figure 7.

Table 6 - Influence of incubation time of the different strains in the presence of 100 ppm cadmium.

Cells / Conditions	100 ppm 3hs	100 ppm 24hs
Saccharomyces cerevisiae S288C	350	880

Saccharomyces cerevisiae 1978	1200	5600
Saccharomyces cerevisiae 2041	4900	8100
Saccharomyces cerevisiae 2464	1600	5400
Starmerela meliponinorum	410	1200
Pichia membranaefaciens	350	4300
Candida cylindracea	730	2800

ND = Not Determined cadmium.

The values were expressed in grams of cadmium incorporated per gram of wet weight of cells.

The data and experimental conditions are taken from Figure 9.

Table 7 - Influence of incubation time and different conditions on cadmium incorporation by yeast exposed to 50 ppm cadmium chloride.

Cells	[1] 50 ppm 3 hrs	[2] 50 ppm 24 hours	[3] 50 ppm 12 hrs	[4] 50 ppm 48 hours Biomass not viable
Saccharomyces cerevisiae W303- WT	ND	ND	450	3400
Saccharomyces cerevisiae S288C	210	1500	720	ND
Saccharomyces cerevisiae 1978	440	1700	440	5600
Saccharomyces cerevisiae 2041	530	1400	2900	4000
Saccharomyces cerevisiae 2464	910	4300	600	5500
Starmerela meliponinorum	160	460	470	6200
Candida cylindracea	440	480	ND	3800
Pichia membranaefaciens	190	1700	ND	3000

ND = Not Determined cadmium.

The values were expressed in grams of cadmium incorporated per gram of wet weight of cells.

The data and experimental conditions for 50 ppm (3 hours and 24 hours) are taken from Figure 9. Data and experimental conditions for 50 ppm, 12 hours have been taken from Figure 10. Data and experimental conditions for 50 ppm, 48 hours (non-viable biomass) are taken from Table 4.

The cells were collected in various situations:

1-   Cells in the exponential growth phase were incubated for 3 hours in the presence of 50 ppm cadmium, after which the cells were collected;

2-   Cells in the exponential growth phase were incubated for 24 hours in the presence of 50 ppm cadmium, after which the cells were collected;

3-   Cells in stationary phase (after 48 hours), then 50 ppm cadmium was added. The cells remained in contact with the metal for 12 hours and were then collected;

4-   Cells were incubated for 12 hours in YPG medium, then autoclaved, after which 50 ppm cadmium was added. The cells remained in this situation for 48 hours, when they were collected.

Table 8 - Influence of incubation time and different conditions on cadmium incorporation by different

yeast strains in the presence of 50 ppm and 100 ppm cadmium chloride.

Cells	100 ppm 24 hours	50 ppm 48 hs
Saccharomyces cerevisiae S288C	880	ND
Saccharomyces cerevisiae 1978	5600	5600
Saccharomyces cerevisiae 2041	8100	4000
Saccharomyces cerevisiae 2464	5400	5500
Starmerela meliponinorum	1200	6200
Candida cylindracea	4300	3800
Pichia membranaefaciens	2800	3000

ND = Not Determined cadmium.

The values were expressed in grams of cadmium incorporated per gram of wet weight of cells.

Data and experimental conditions for 100 ppm (24 hours) are taken from Figure 9. Data and experimental conditions for 50 ppm 48 hours (Biomass not viable) are taken from Table 4.

100 ppm 24 hours 50 ppm 48 hs

6 DISCUSSION

In this work we studied various biological, physiological and chemical parameters in order to select yeasts that could be efficient at capturing cadmium from aqueous media.

The biological parameters studied in yeast were: cell growth in liquid culture, colony formation in solid medium, cadmium tolerance, trehalose levels after exposure to cadmium, induction of metal resistance, cadmium uptake by cells. The physiological parameters studied in the yeast were: different phases of cell growth, cell density of the inoculum, capacity for metal uptake by live and dead cells. The chemical parameters studied were: different concentrations of cadmium, chemical form of cadmium, time of exposure to the metal.

The study was first carried out on a laboratory yeast strain. The aim was to have reference values to guide the search for an efficient yeast for metal uptake. Subsequently, we used yeasts used in the fermentation of cachaça produced in the Minas Gerais region and yeasts isolated from regional ecological niches.

Cadmium was chosen because it is a toxic metal that is associated with various health problems (ATSDR, 1997; WHO, 1992). Cadmium is present in natural waters due to industrial effluent discharges, mainly in electroplating, pigment production, welding, electronic equipment, lubricants and photographic accessories. The burning of fossil fuels is also a source of cadmium release into the environment (ILO, 1998; MEDITEXT, 2000).

In humans, cadmium can have a chronic effect, as it concentrates in the kidneys, liver, pancreas and thyroid, and has an acute effect, where a single dose of 9.0 grams can lead to death (LÓPEZ-ALONSO et al, 2000).

Cadmium is an element with no biological function, with the only known exception of the carbonic anhydrase of the alga *Thalassiosira weissflogi* (LANE et al, 2005), which is dependent on the metal. It is toxic even at low concentrations. All the strains studied, both laboratory strains and strains isolated from cachaça fermentation or from the region, were susceptible to cadmium. Figure 1 shows that the presence of 2.0 ppm cadmium chloride interfered with the growth of the laboratory strain of *Saccharomyces cerevisiae*. At all the concentrations used, above 2.0 ppm cadmium chloride, there was a significant inhibition of growth over the 28-hour period during which the study was carried out. The same inhibitory effect on growth was also seen for yeasts isolated from cachaça fermentation and from the region (Figure 6).

GHARIEB (2001), working with the fungus *Fusarium sp.*, showed that growth inhibition increased as the concentration of cadmium in the medium increased. According to the literature, the inhibitory effect of cadmium on cell growth is due to several factors, all of which are linked to the blocking of functional groups of important molecules such as: enzymes, polynucleotides, ion or nutrient transport systems and the replacement of essential ions in active sites (GHARIEB, 2001).

Cadmium induces the formation of free radicals, not through the Fenton reaction, but as a result of its reaction with thio groups. Proteomic studies carried out on the yeast *Saccharomyces cerevisiae* show that the expression of many proteins is altered in response to cadmium, particularly enzymes responsible for sulfate assimilation, heat shock proteins, oxidative stress enzymes, proteases and carbohydrate metabolism enzymes. The most obvious changes in proteases and carbohydrate metabolism enzymes are the appearance of low-sulphur isoforms which are synthesized to take the place of pre-existing enzymes. In the absence of cadmium, the amount of sulphate incorporated into these proteins is 79%, but in the presence of cadmium, it drops to 19%. Thus, the amino acids that contain sulphate are directed towards the glutathione (GSH) synthesis pathway. The fact that these new enzymes have fewer amino acids with sulphate makes them less susceptible to the deleterious effect of cadmium, as it will now bind with less affinity to them and remain free (MENDOZA-CÓZATL et al, 2004). Cd-resistant strains *of Saccharomyces cerevisiae* are protected from cadmium toxicity by the production of large quantities of metallothionein and phytochelatins that bind the metal (GHARIEB, 2001).

It is believed that the toxic effect of cadmium is also a consequence of structural damage to the plasma membrane, possibly by binding to organic components such as sulfhydryl groups and by altering the permeability of essential ions such as Na^+ , K^+ , Mg^{+2} and Ca^{+2} . These ions are present in relatively high concentrations in biological systems and are essential for various cellular functions including the formation of charges and concentration gradients across membranes which are used in transport processes, osmotic responses, maintenance of cytoplasmic pH, stabilization of ribosomes, nucleic acids and activating enzymes that synthesize DNA, RNA and proteins (GHARIEB, 2001). Cadmium can be transported from the extracellular environment to the inside of the cell in *Saccharomyces cerevisiae* via the zinc transporter, thanks to the similarity in chemical coordination between the two metals (GOMES et al, 2002).

Finally, we can consider that the toxicity of metals in living organisms involves oxidative or genotoxic mechanisms. These oxidative or genotoxic mechanisms are based on the physical and chemical properties of metals. Depending on the metal, there is a molecular mechanism. In practice, up to three molecular mechanisms of heavy metal toxicity are proposed: production of reactive species by autooxidation and the Fenton reaction (Fe, Cu), blocking of essential functional groups in biomolecules (Cd, Hg), and displacement of metal ions from biomolecules (for example, Cd replacing Ca^{+2} or, Zn^{+2}) (GOMES et al, 2002; ALKORTA et al, 2004). Studies indicate that the genotoxic action of cadmium is linked to the inactivation of repair systems, thus inducing hypermutability (MCMURARY & TAINER, 2003). In general, DNA repair systems involve: reversal of direct damage, base excision, nucleotide excision, repair of double strand breaks, and mismatch repair. In humans and in the yeast *Saccharomyces cerevisiae*, cadmium's action occurs because the metal specifically inactivates the mismatch repair system, so it is not a direct action of the metal on DNA but rather inhibits a specific type of DNA repair (MCMURARY & TAINER, 2003).

Cadmium as a toxic (ATSDR, 1997; WHO, 1992) and non-essential metal is a strong stressor (BRENNAN & SCHIESTL, 1996). Cells respond to the most varied stresses possible in a few ways: by inhibiting all genes that are not essential for the new situation and by strongly stimulating genes that are somehow involved in the stress in question or that can protect the cell in some way (LINDQUIST & CRAIG, 1988). Thus, our aim was to see whether cells respond to the presence of cadmium by synthesizing trehalose. The choice of determining trehalose levels was due to the fact that this carbohydrate is an unspecific marker of stress and its levels are generally found to be increased in a wide variety of situations perceived as harmful to yeast cells (VAN

LAERE, 1989). Figure 2 shows that an increase in cadmium concentrations leads to an increase in the levels of trehalose accumulated by the laboratory yeast strain. The same trehalose synthesis response can be seen in the strains isolated from cachaça fermentation and strains isolated from the region (Figure 7), however, the results in this protocol are not as clear as those observed in the laboratory strain (Figure 2), Given our interest in carrying out the experiment in a situation of high cadmium concentration (500 ppm), our aim is to select a yeast that can capture cadmium in highly polluted industrial effluents.

The strains *Saccharomyces cerevisiae* 1011, *Saccharomyces cerevisiae* 2057, *Saccharomyces cerevisiae* 2464, *Starmerela meliponinorum, Candida cylindracea, Candida sorboxylosa, Pichia guilliermondii, Pichia kluyvera* and *Pichia membranaefaciens* were able to accumulate more trehalose compared to the control when incubated in the presence of 500 ppm $CdCl_2$. Obviously, this is a lethal concentration (Table 2), however, in the 3 hours that these cells remained in contact with this high concentration of metal, they managed to accumulate high levels of trehalose. Cells with high levels of trehalose are generally more resistant (LILLIE & PRINGLE, 1980).

The first function attributed to trehalose was as a reserve carbohydrate (ELBEIN, 1974). Later, correlations were made between trehalose content and resistance to various stress conditions. Thus, the disaccharide trehalose protects against desiccation stress (D'AMORE et al, 1991; GADD et al, 1987; HOTTINGER et al, 1987), high temperatures (HOTTINGER et al, 1987; NEVES & FRANÇOIS, 1992), freezing (LEWIS et al, 1995), increased hydrostatic pressure (FERNANDES et al, 2001), oxidizing agents such as ethanol, copper sulphate, hydrogen peroxide (LUCERO et al, 2000; MANSURE et al, 1994; RIBEIRO et al, 1999), osmotic stress (HOUNSA et al, 1998) and heavy metals (ATTFIELD, 1987).

The protective function of trehalose seems to be clear only in cases of desiccation and heat shock. During the process of desiccation or dehydration, biological membranes lose their structural and functional integrity. It is believed that trehalose interacts with membrane phospholipids, forming hydrogen bridges with their OH groups and phosphate groups. These hydrogen bridges could take the place of water around the phospholipids, thus stabilizing dehydrated membranes (CROWE et al, 1991; DE ARAUJO et al, 1991; ELEUTHERIO et al, 1993). In the case of temperature rise (heat shock), the presence of trehalose prevents the aggregation of proteins denatured by the thermal effect, thus maintaining the active conformational state of the proteins (SINGER & LINDQUIST, 1998). In all other stresses in which trehalose accumulates, its function is unknown.

Despite all the above considerations, we should point out that in our experiments we did not see a marked protective role for trehalose (figures 5A and 5B). In these experiments (figures 5A and 5B), we induced the synthesis of trehalose with a concentration of 2.5 ppm cadmium chloride and then added a further 50 ppm cadmium chloride to the cells which had increased levels of trehalose. We did not find that the presence of high levels of trehalose was an additional advantage for the cells to continue growing in this toxic condition. This lack of correlation between increased levels of trehalose and protection has already been verified by other authors (ALEXANDRE et al, 1998), however, to date, there is no clear understanding of why the cell synthesizes so much trehalose if it is not going to protect it. Of course, we must always consider that various gene circuits for detecting stresses and activating genes are interconnected and overlapping, so they are not highly specific (BANUETT, 1998). Another relevant consideration is that periods of reduced growth are associated with high levels of trehalose. Thus, cells in a stationary phase, or cells unable to grow due to a deficiency in essential nutrients such as carbon, nitrogen, sulphur and phosphate (LILLIE & PRINGLE, 1980) or dormant cells such as spores (INOUE & SHIMODA, 1981), conidia (HANKS & SUSSMAN, 1969) and ascospores (SUSSMAN & LINGAPPA, 1959), have high levels of trehalose. In this case, it is clear that there is a reduction in growth (Figure 1 and Figure 6) and, perhaps, it is possible to speculate that this difficulty in continuing the normal cell cycle is what signals the synthesis of trehalose rather than the presence of a stress agent. This possibility has already been proposed by THEVELEIN (1996), although the author was not dealing with the specific case of cadmium, but with the global role that trehalose can play in the life cycle of the yeast *Saccharomyces cerevisiae*.

The presence of heavy metals affects the metabolic activity of fungal and yeast cultures and can affect the commercial fermentation process. The effect of the presence of heavy metals on commercial fermentation processes has generated interest in correlating the behavior of fungi in their presence. The results of these studies led to the idea that fungi and yeasts could be used to remove heavy metals from industrial effluents.

Due to their immutable nature, metals are a group of pollutants that receive a lot of attention. In fact, many metals are essential for biological systems and must be present in certain concentrations. If the concentration of the metal increases beyond certain parameters, the metals can act in a deleterious way, blocking essential functional groups, displacing other metals or modifying the conformation of biological molecules.

As a result of human activity such as mining, electroplating, the production of fuels, energy, fertilizers and the application of pesticides, metal pollution is one of the major environmental and health problems of today. Large proportions of heavy metals have been released into the environment as industrial waste or industrial effluents that reach bodies of water, thus contributing to increased pollution in aquatic systems (BABICH & STOTZKY, 1980). More restrictive environmental legislation, ecological problems and the high cost of conventional technologies for treating effluents containing heavy metals have resulted in a search for unconventional technologies. Many yeasts, microorganisms and algae are known for their ability to concentrate heavy metals from aqueous solutions, accumulating them inside or on the surface of their cell structures (BRADY & DUNCAN, 1994; VOLESKY & MAY-PHILLIPS, 1995; GOMES, et al, 2002; GADD, 1986; HUANG et al, 1988a; KAPOOR & VIRARAGHAVAN, 1995; GAVRILESCU, 2004; VOLESKY, 1990a). This work is part of the search for new alternatives to reduce the concentration of metals that can pollute the environment.

Fungi and yeasts are easy to grow and produce a lot of biomass. They are widely used in a variety of industrial fermentation processes. This biomass has a low cost because it is the end product of a fermentation process and would have to be discarded. The possibility of using biomass for something other than disposal could generate new dividends for the industry. Our interest in using selected yeasts from the fermentation of cachaça is due to the fact that the state of Minas Gerais has a large number of stills producing the drink. In 1995, according to a SEBRAE survey, the number of stills in a regular situation in the state was 8,466. It is known that the sector has a high level of informality, so it is possible that the number shown in the SEBRAE survey was underestimated (SEBRAE, 2004).

On the other hand, we were also interested in using yeasts isolated from the region because, if there is an interest in cultivating them for biomass production, their cultivation is usually simple, done with low-cost technology and using inexpensive culture media (KAPPOR & VIRARAGHAVAN, 1995). It is important to note that microorganisms isolated from a region are highly adapted to that region, so variations due to seasonality, such as temperature, amount of water available and humidity, have no influence on the microorganism, since it is perfectly adapted to the conditions prevailing in the region (ZUZUARREGUI & DEL OLMO, 2004).

We found that cells from the W303-WT laboratory strain of *Saccharomyces cerevisiae* are able to remove cadmium from the extracellular medium and accumulate the metal (Figure 8). Cadmium uptake is dependent on the concentration of the metal present in the medium and the length of time the cells are in contact with this medium. Increasing the concentration and exposure time to cadmium increases the amount of cadmium incorporated by the cells. The same result was seen in cells isolated from cachaça fermentation and from the region (Figure 9). Comparing the results shown in Figure 9 with the results of cadmium incorporation by the cells of the laboratory strain in Figure 8, it can be seen that the cells isolated from cachaça fermentation and from the region incorporate much more cadmium than the laboratory strain. To illustrate this, the yeast strain *Saccharomyces cerevisiae* W303-WT (laboratory strain) incorporated 820 g of cadmium when exposed to 100 ppm of metal for 20 hours (Figure 8) and the cells of the yeast strain *Saccharomyces cerevisiae* 2041 incorporated 8100 g of cadmium when exposed to 100 ppm of metal for 24 hours (Figure 9).

It can be seen that the yeast strain *Starmerela meliponinorum* had the lowest cadmium uptake (Figure 9, 100 ppm cadmium 24 hours and compare with the other strains in the same graph), however, this strain was the only one that grew in the presence of a concentration of 100 ppm cadmium chloride (Table 2). This phenomenon may be due to the fact that the strain that is most resistant to a metal is not always the one with the highest metal uptake. The cells may be using a metal-extruding mechanism, which allows them to grow in a high concentration of metal, but not to incorporate it in large quantities (SHIEAISHI et al, 2000). Results from other authors (GHARIEB, 2001) suggest that cadmium efflux is an inducible mechanism that begins in the presence of a cadmium gradient across the cell membrane. This mechanism may be the induction of an active exporter system that limits the intracellular concentration of the metal. This mechanism occurs in cadmium-resistant mutants of *Saccharomyces cerevisiae* (SHIEAISHI et al, 2000). It is part of one of the four mechanisms proposed for a cell to be resistant to a metal (SHIEAISHI et al, 2000). The other three mechanisms are: repression of a metal-specific transporter gene, induction of specific proteins that can bind the metal, such as phytochelatins, glutathione and metallothioneins, and compartmentalization in vacuoles, limiting the cytoplasmic concentration of the ion (SHIEAISHI et al, 2000).

Figure 9 also shows that the laboratory strain *Saccharomyces cerevisiae* S288C also incorporated less cadmium than yeasts isolated from nature (compare 50 ppm or 100 ppm cadmium over 24 hours in all the strains shown). This result is extremely supportive of the initial hypothesis that yeast strains isolated directly from nature would perform better at capturing the metal than laboratory yeast strains. Since strains isolated directly from nature should be better adapted to the fluctuations and stresses of the environment.

In fact, the results obtained by VIANNA on various yeast strains isolated from cachaça fermentation (VIANNA, 2003) corroborate the claim that resistance is greater in microorganisms isolated directly from nature. In this work, VIANNA (2003) determined the synthesis of heat shock proteins in various simulations of stresses that occur during fermentation for the production of cachaça. These strains showed constitutive synthesis, i.e. in the absence of stress, of the heat shock proteins hsp 70 and hsp 104. Among the various strains the author used were also the yeasts *Saccharomyces cerevisiae* 1011 and 2464, used in the present work. The cellular response to stress is evolutionarily conserved in all living organisms and the major role is attributed to heat shock proteins (Hsps) and other molecules that confer protection against stress, such as trehalose. The molecular response induced by the cell determines whether the organism adapts, survives or whether the injury will be repaired or lead to death (ALOYSIUS, 1999). Normally, the expression of heat shock proteins occurs in all stresses and seems to be linked to the maintenance of vital cellular functions in the face of a harmful stimulus (ALOYSIUS, 1999). The constitutive expression of hsp 70 and hsp 104 in various yeast strains could indicate that yeasts isolated from cachaça fermentation, and perhaps from our habitat, could be more resistant to stresses in general. In addition to constitutive hsp synthesis, the *Saccharomyces cerevisiae* strains isolated from cachaça fermentation used by VIANA show high levels of trehalose in the absence of stress when compared to laboratory strains (VIANNA, 2003).

According to the literature, the uptake of metals by living cells depends on the contact time (Figure 8 and Figure 9), pH of the solution, culture conditions, initial metal concentration (Figure 8 and Figure 9) and cell concentrations (Table 3), (KAPPOR & VIRARAGHAVAN, 1995).

To study the effect of cell density on metal uptake, we carried out the experiment shown in Table 3. The incubation of cells from the laboratory strain (*Saccharomyces cerevisiae* S288C) was carried out keeping the volume, exposure time and cadmium concentration fixed, with only the mass of cells placed in the medium varying. It can be seen that, despite the great difference in the cell mass used, the levels of cadmium incorporated were basically the same. This indicates that larger amounts of cell mass do not favor the incorporation of cadmium by the cells. Low cadmium uptake in biomass with high cell density has already been observed in several fungal strains, such as: *Aspergillus oryzae, Aspergillus niger, Mucor racemosus, Penicillium chrysogenum, Trichoderma viride* (KAPOOR & VIRARAGHAVAN, 1995). KAPPOR & VIRARAGHAVAN (1995) attributed this phenomenon to electrostatic interactions on the surface of the cells. The presence of a large number of cells implies that the distance between them has decreased, allowing rapid aggregation and the formation of lumps, reducing the surface area available for contact with the metal (KIFF & LITTLE, 1986).

To check the influence of culture conditions on metal uptake, the experiment shown in Figure 10 was carried out. In this experimental protocol, the cells were grown for 48 hours in YPG medium until they reached stationary phase (verified by the depletion of glucose with an indicator strip) when 50 ppm of cadmium chloride was added for 12 hours. It can be seen that the incorporation of cadmium was lower when compared to cells grown in the presence of 50 ppm cadmium chloride (Figure 9, 50 ppm cadmium for 24 hours). The accumulation of metals by growing cells varies with the age of the cells. One possible explanation for this is that changes in cell morphology lead to chemical changes on the cell surface, favoring greater metal uptake. Maximum metal uptake occurs during the lag phase or in the early stages of growth and declines when the culture reaches the stationary phase (KAPOOR & VIRARAGHAVAN, 1995). When using cells in the stationary phase of growth, the accumulation of metals in these cells is generally very rapid and reaches equilibrium between 60 and 120 minutes, however, 90% of this uptake occurs within 10 minutes (KAPOOR & VIRARAGHAVAN, 1995). Biosorption of metal ions occurs through surface binding, including ion exchange, and complexation with functional groups present on the cell surface. It is believed that several functional groups may be involved in binding metals. These functional groups include carboxylic, sulfhydryl, amine and phosphate groups (KAPOOR & VIRARAGHAVAN, 1995).

Yeasts, algae, fungi and bacteria have cell walls with organic acids and functional groups capable of adsorbing protons and metal cations from an aqueous solution. The uptake of heavy metals by biomass can take place by an active mechanism (dependent on the metabolic activity of the cell) known as bioaccumulation and by a passive mechanism (sorption and/or complexation) known as biosorption (VOLESKY & MAY-PHILLIPS, 1995; VOLESKY, 1990a). Biosorption is the passive absorption of metal ions on the cell surface and bioaccumulation is when the metal ion is transported through the cell to its interior (BLACKWELL & TOBIN, 1999). To check whether the mechanism of action used by the cells to capture metal was bioaccumulation or biosorption, the experiment shown in Table 4 was carried out. In this experimental protocol, the cells were grown to logarithmic phase in YPG medium. These cultures were then placed in an autoclave (15 minutes at a temperature of 121^0 C, under high pressure) to kill the cells. Subsequently, 50 ppm cadmium chloride was added to these cells. The cells remained in contact with the metal for 48 hours, when they were collected to determine the amount of cadmium incorporated. Table 4 shows that dead cells incorporate a greater amount of metal than live cells (Table 7). Our data indicate that the biosorption mechanism (Table 4) is efficient for cadmium uptake. According to the literature, the biosorption of metals by non-viable cells is possible thanks to the presence of functional groups on the external surface that allow the metal to bind (KAPOOR & VIRARAGHAVAN, 1995).

Table 8 compares the results obtained during bioaccumulation with biosorption. The difference between accumulation by one mechanism or the other is not that different, although it is a difficult comparison to make since the concentrations used and the exposure times were different. However, the use of dead cells has many advantages, as will be seen below.

Dead biomass can be obtained from industrial sources, which use yeast cells for the fermentation process of food and beverages. The cells can be killed by raising the temperature (SIEGEL et al, 1986; SIEGEL et al, 1987; GALUN et al, 1983; TOWNSLEY et al, 1986), by autoclaving (HUANG et al, 1988a; TOBIN et al, 1984) or chemical substances such as acids, alkalis, detergents and (AZAB & PETERSON, 1989; BRADY et al, 1994; RAO et al, 1993; HUANG et al, 1988a; MUZZARELLI et al, 1980a; WALES & SAGAR, 1990; TSEZOS & VOLESKY, 1981; ROSS & TOWNSLEY, 1986; GADD et al, 1988), other organic chemical substances such as formaldehyde (HUANG et al, 1988b) or even mechanical methods to rupture the cells. All of these methods can, in some way, interfere with the metal's ability to bind to the groups present in the wall, favoring or hindering this binding. The method of autoclaving to kill the cells was chosen so as not to generate more chemical residues. Since, if I had opted for chemical treatments to kill the cells, I would have introduced the use of this chemical, which would then be a waste product that would somehow have to be treated to neutralize it.

The use of non-viable biomass for metal uptake offers some advantages over live cells. Systems using live cells are more sensitive to the toxic effects of metal concentration and conditions such as pH and temperature. Live cells require a constant supply of nutrients, which certainly has an influence on operating costs.

Yeasts, algae, fungi and bacteria can remove heavy metals and radionuclides from aqueous solutions in substantial quantities (ADAMIS et al, 2003; ALOYSIUS et al, 1999; GADD, 1986).

The levels of metals and organic compounds that can be present in water depend on what the water will be used for. In Brazil, these limits are set by CONAMA Ordinance No.⁰ 357 (2005). In general, industrial effluents show levels of metals above those permitted for disposal. As an example, we can cite the work of BENEDETTO et al (2005) which characterized the industrial effluent of Companhia Paraibuna de Metais, which has a flow rate of over 100 m^3 /h, located in the state of Minas Gerais. The authors determined the following composition: Zn=108mg/L, Ca=208mg/L, Mg=105mg/L, Mn=193mg/L, Fe=2.7mg/L, Cd=36mg/L, Pb= 1.92mg/L. All the metals present are above the levels permitted by CONAMA resolution 357. Therefore, conventional treatments such as precipitation with different chemical compounds or removal by commercial resins are required for disposal. Below are the adsorption results obtained by the authors when they tested various commercial resins or absorbents, using 100 ml of this effluent and 0.5 grams of resin (BENEDETTO et al2005):

Table 9 - Removal of metals from the effluent of Companhia Paraibuna de Metais using various absorbent resins.

Adsorbent	Synthesized	Type	Functional group	% Adsorption						
				Zn	Ca	Mg	Mn	Fe	Cd	Pb
IR 748	ROHM & HAAS	Chelating	Iminodiacetic acid	26,5	<0.5	1,0	38,6	77,9	26,3	86,2
IRA-120	HOHM & HAAS	Cationic strong	Sulphonic acid	39,2	25,0	34,3	<0.5	18,9	35,5	70,5
S 950	PUROLITE	Chelating	Aminophosphoric Acid	35,9	<0.5	6,7	31,6	>99	22,9	83,8
TP 207	BAYER	Chelating	Iminodiacetic acid	13,0	<0.5	<0.5	39,4	>96	1,7	41,7
VPOC 1026	BAYER	Cationic strong	DEHPA	53,7	9,1	9,5	56,0	55,2	7,5	27,6
C 249	SYBRON	Chelating	Iminodiacetic	49,1	64,3	44,8	68,9	90,4	46,4	85,9

	CHEMICALS		acid								
ZEOLITE	PETROBRAS	-	-	1,9	15,4	<0.5	40,0	<0.5	<0.5	22,9
COAL	BONECHAR BR	-	-	-	2,3	<0.5	0.5	7,2	>99	41,9	>99

It can be seen in this table (BENEDETTO et al, 2005) that, for cadmium, the best removal was achieved with the C 249 resin which, under the conditions used by the authors, removed 3.34 mg/gram of resin. The values obtained using yeasts are close to, or better than, those presented by this resin. See table 4, where 50 ppm of cadmium chloride was used (the value closest to the initial concentration of the effluent), and the retention of cadmium by the yeasts was higher than that of this resin, with the exception of the *Pichia kluyvera* strain. Our results are presented in wet weight, which means that when expressed in dry weight, this value will be higher.

This result shows that yeasts have great potential for this cadmium removal. Of course, we must emphasize that our results were obtained using a single metal, while the real effluent contains seven metals. It is likely that interference between the various metals will reduce the incorporation of a given metal by the yeast. Evaluating the data obtained in this work, the use of yeast as a biosorbent material seems feasible.

7 CONCLUSION

- Cadmium, even in low concentrations, interferes with the growth of various yeast cell strains, including strains isolated from cachaça fermentation and the regional environment.

- The presence of cadmium induces trehalose synthesis.

- The presence of high levels of trehalose induced by cadmium in the laboratory strain (W303-WT) did not provide much protection to the cells challenged with the metal.

- The presentation of cadmium in different chemical forms had no influence on the amount of cadmium incorporated by *Saccharomyces cerevisiae* W303-WT cells (laboratory strain).

- All the cells of the strain used in this work (with the exception of the *Saccharomyces cerevisiae* laboratory strain) tolerated concentrations of up to 50 ppm of cadmium chloride.

- The yeast strain *Starmerela meliponinorum* grew in the presence of 100 ppm cadmium chloride, but, using viable cells, showed low cadmium incorporation compared to the other yeast strains.

- Cadmium incorporation by viable cells depends on the time of exposure to the metal and the cadmium concentration present in the medium. Time and higher concentrations favor the incorporation of cadmium by yeasts.

- Cadmium incorporation depends on the life stage of the yeast. Cells in logarithmic phase incorporate more cadmium than cells in stationary phase.

- Increasing masses do not favor the incorporation of cadmium into cells.

- Yeast strains isolated from the fermentation of cachaça and from the region incorporate higher amounts of cadmium when compared to laboratory strains.

- Dead yeasts incorporate cadmium from the extracellular medium.

- There were no significant differences in cadmium incorporation between the strains studied.

- The results suggest that the use of an isolated yeast strain will not lead to greater incorporation of cadmium.

The results suggest that the best way to incorporate cadmium from contaminated effluents is to use biomass made up of different genera and species of dead yeasts.

8 BIBLIOGRAPHICAL REFERENCES

ADAMIS, P. D. B.; PANEK, A. D.; LEITE, S. G. F.; ELEUTHERIO, E. C. A. Factors involved with cadmium absorption by a wild-type strain of *Saccharomyces cerevisiae*. **Brazilian Journal of Microbiology**, v.34, p.55-60, 2003.

ALBERTINI, S. **Cadmium adsorption isotherms by *Saccharomyces cerevisiae*.** 54p. Dissertation (master's degree) - Escola Superior de Agricultura "Luiz de Queiroz", Universidade de Sao Paulo. Piracicaba, 1999.

ALEXANDRE, H.; PLOURDE, L.; CHARPENTIER, C.; FRANCOIS, J. Lack of correlation between trehalose accumulation, cell viability and intracellular acidification as induced by various stresses in *Saccharomyces cerevisiae*. **Microbiology**, v.144, n.4, p.1103-1111, 1998

ALKORTA, I.; HERNANDES-ALLICA, J.; BECERRRIL, J. M.; AMEZAGA, I.; ALBIZU, I.; GARBISU, C. Recent findings on the phytoremediation of soils contaminated with environmentally toxic heavy metals and metalloids such as zinc, cadmium, lead and arsenic.

Reviews in Environmental Sciences and Biotechnology, v.3, p.71-90, 2004.

ALOYSIUS, R.; KARIM, M. I. A.; ARIFF, A. B. The mechanism of cadmium removal from aqueous solutions by nonmetabolizing free and immobilized live biomass from *Rhizopus oligosporus*. **World Journal of Microbiology & Biotechnology**, v.15, p.571-578, 1999.

ARANHA, S.; NISHIKAWWA, A. M.; TAKA, T.; SALIONE, E. M. C. Cadmium and lead levels in bovine liver and kidneys. **Revista Instituto Adolfo Lutz**, v.54(1), p. 16-20, 1994.

ASSMANN, S.; SIGLER, K.; HOFER, M. Cd^{+2} induced damage to yeast plasma membrane and its alleviation by Zn^{+2} : studies on *Schizosaccharomyces pombe* and reconstituted plasma membrane vesicles. **Archive of Microbiology**, v.165, n.4, p.279-282, 1996.

ATSDR Agency for Toxic Substances and Disease Registry. **Toxicological Profile for Cadmium**, Atlanta: ATSDR, 1997. p.347

ATTFIELD, P. V. trehalose accumulates in *Saccharomyces cerevisiae* during exposure to agents that induce heat shock response. **FEBS Letters**, v.225 n.1-2, p.259-263,1987.

AZAB, M. S.; PETERSON, P. J. The removal of cadmium from water by the use of biological sorbents. **Water Science Technology**, v.21, p.1705-1706, 1989.

BABICH, H.; STOTZKY, G. Environmental factors that influence the toxicity of heavy metals and gaseous pollutants to microoganisms. **CRC Critical Reviews Microbiology**, v.8, n.7, p.99-145, 1980.

BABICH, H.; STOTZKY, G. Influence of chemical speciation on the toxicity of heavy metals to the microbiota. In: NRIAGU, J.O. (Ed.) **Aquatic Toxicology**, New York: Wiley & Sons, p.1-46. 1983.

BELL, W.; SUN, W.; HOHMANN, S.; WERA, S.; REINDERS, A. DE VIRGILIO, C.; WIEMKEN, A.; THEVELEIN, J. M. Composition and functional analysis of the *Saccharomyces cerevisiae* trehalose synthase complex. **Journal of Biological Chemistry,** v.273, p.33311-33319, 1998.

BENAROUDJ, N.; LEE, D. H.; GOLDBERG, A. L. Trehalose accumulation during cellular stress protects cells and cellular proteins from damage by oxygen radicals. **Journal of Biological Chemistry,** v. 276, n.26, p.24261-24267, 2001.

BENEDETTO, J. S.; MORAIS, C. A.; MELO, F. S. D. Zinc And Heavy Metals Recovery/Removal From Industrial Effluent By Ion Exchange. In: ICHMET

INTERNATIONAL CONFERENCE ON HEAVY METALS IN THE ENVIROMENT XIII, 2005, Rio de Janeiro, RJ. **Proceedings...** Rio de Janeiro: Mineral Technology Center, 2005. CD-ROM.

BANUETT, F. Signalling in the yeasts: an informational cascade with links to the filamentous fungi. **Microbiology and Molecular Biology Reviews,** v.62, n.2, p.249-274, 1998.

BEVERIDGE, T. J. Role of cellular desing in bacterial metal accumulation and mineralization. **Annual Review Microbiology,** v.43, n.3, p.147-171, 1989.

BIRCH, L.; BACHOFEN, R. Complexing agents from microorganisms. **Experientia,** v.46, n.7, p.827-834, 1990.

BLACKWELL, K. J.; TOBIN, J. M. Cadmium accumulation and its effects on intracellular ion pools in a brewing strain of *Saccharomyces cerevisiae*. **Journal of Industrial Microbiology & Biotechnology,** v.23, p.204-208, 1999.

BRADY, D., STOLL, A.; DUCAN, J. R. Biosorption of heavy metal cations by non-viable yeast biomass. **Environmental Technology,** v. 15, p.429-438, 1994.

BRADY, D.; DUNCAN, J. R. Bioaccumulation of metal cation by *Saccharomyces cerevisiae*.

Applied Microbiology and Biotechnology, v.41, p.149-159, 1994.

BRENNAN, R. J.; SCHIESTL, R. H. Cadmium is an inducer of oxidative stress in yeast.

Mutation Research, v.356, n.1, p.171-178, 1996.

BROCK, T. D.; SMITH, D.W.; MADIGAN, M. T. **Biology of Microorganisms,** 4ed. Prentice Hall: New Jersey, 1984. 909p.

CABIB, E.; LELOIR, L. F. The biosynthesis of trehalose phosphate. **Journal of Biological Chemistry,** v.231, p. 259-275, 1958.

CETESB. Environmental Sanitation Technology Company. **Report on the establishment of guideline values for soil and groundwater.** Sao Paulo: CETESB, 2001.

NATIONAL RESEARCH COUNCIL. Committee on in Situ Bioremediation. **Situ bioremediation**: when does it work? Washington: National Academies Press, 1993. Available at: <http://www.nap.edu/catalog/2131.html>. Accessed on: August 19, 2005.

CONAMA. National Environment Council. **CONAMA Resolution 357**: establishes the classification of waters and the required quality levels. Available at: <http://www.mma.gov/port/conama/res/res86/res2086.html>. Accessed on: June 13, 2005.

COSSICH, E. S. **Biosorption of Chromium (III) by the Biomass of the Marine Algae *Sargassum* sp**. 147p. Campinas: School of Chemical Engineering, State University of Campinas, Thesis (Doctorate), 2000.

COUTINHO, C. C.; SILVA, J. T.; PANEK, A. D. Trehalose activity and its regulation during growth of *Saccharomyces cerevisiae*. **Biochemistry International,** v.26, p.521-530, 1992.

CROWE, J. H.; CROWE, L. M.; CHAPMAN, D. Preservation of membranes in anhydrobiotic organisms: the role of trehalose. **Science,** v.223, p. 701-703, 1984.

CROWE, J. H.; PANEK, A. D.; CROWE, L. M.; PANEK, A. C.; DE ARAUJO, P. S. Trehalose transport in yeast cells. **Biochemistry International,** v.24, p.721-730, 1991.

D'AMORE, T.; CRUMPLEN, R.; STEWART, G. G. The involvement of trehalose in stress tolerance. **Journal Industrial Microbiology,** v.7, p.191-196, 1991.

DE ARAUJO, P. S.; PANEK, A. C.; CROWE, J. H.; CROWE, L. M.; PANEK, A. D. Trehalose-transporting membrane vesicles from yeasts. **Biochemistry International**, v.24, p.731-737, 1991.

DE SOETE, D.; GIJBELS, R.; HOSTE, J. **Neutron Activation Analysis,** New York: John Wiley, p.177, 1972.

DE VIRGILIO, C.; BURCKERT, N.; BELL, W.; JENO, P.; BOLLER, T.; WIEMKEN. A. Disruption of *TPS2*, the gene encoding the 100kDa subunit of the trehalose-6-phosphate synthase / phosphatase complex in *Saccharomyces cerevisiae*, causes accumulation of trehalose-6-phosphate and loss of trehalose-6-phosphate phosphatase activity. **European Journal of Biochemistry,** v.212, p.315-323, 1993.

DE VIRGILIO, C.; HOTTIGER, T.; BOLLER, T.; WIEMKEN. A. The role of trehalose synthesis for the acquisition of thermotolerance in yeasts. I. Genetic evidence that trehalose is a thermoprotectant. **European Journal of Biochemistry,** v.219, p.179-186, 1994.

DEL RIO, D. T. **Cadmium biosorption by *Saccharomyces cerevisiae* yeasts**. 66p. Master's dissertation - Escola Superior de Agricultura "Luiz de Queiroz" da Universidade de Sao Paulo. Piracicaba, 2004.

DONMEZ, G.; AKSU, Z. The effect of copper (ll) ions on the growth and bioaccumulation properties of some yeasts. **Process Biochemistry**, v.35, p.135-142, 1999.

EHMANN, W. D.; VANCE, D. E. **Radiochemichemistry and Nuclear Method of Analysis,** New York: John Wiley, 1991, 531 p. (series: Chemical Analysis).

ELEUTHERIO, E. C. A.; DE ARAÙJO, P. S.; PANEK, A. D. Role of the trehalose carrier in dehydration resistance of *Saccharomyces cerevisiae*. **Biochimica et Biophysica Acta,** v.1156, p.263-266, 1993.

ELBEIN, A. D. The metabolism of alpha,alpha-trehalose. **Advances in Carbohydrate Chemistry and Biochemistry,** v.30, p.227-256, 1974.

ELLIOT, B.; HALTIWANGER, R. S.; FUTCHER, B. Synergy between trehalose and Hsp 104 for thermotolerance in *Saccharomyces cerevisiae*. **Genetics,** v.144, p.923-933, 1996.

ERASO P; MARTiNEZ-BURGOS M.; FALCÓN-PÉREZ. J. M.; PORTILLO F.; MAZÓN M. J. Ycf1-dependent cadmiun detoxification by yeast requires phosphorylation of residues Ser908 and Thr911. **FEBS Letters,** v.577, p.322-326, 2004).

FERNANDES, P. M. B.; FARINA, M.; KURTENBACH, E. Effect of hydrostatic pressure on the morphology and ultrastructure of wild-type and trehalose synthase mutant cells of *Saccharomyces cerevisiae*. **Letters in Applied Microbiology,** v.32, n.1, p.42-6, 2001.

FRIBERG, L.; PISCATOR, M.; NORDBERG, G. F.; KJELLSTROM, T. **Cadmium in the Environment,** 2 ed. Cleveland, OH: CRC Press, 1974. 94p.

GADD, G. M. The uptake of heavy metals by fungi and yests: the chemistry and physiology of the process and applications for biotechnology. In: ECCLES, H; HUNT, S. **Immobilization of Ions by Bio-Sorption,** London: Ellis Horwood Limited Publishers, 1986. Chap. 5.2, p.135-147.

GADD, G. M.; CHALMERS, K.; REED, R. H. The role of trehalose in dehydration resistance of *Saccharomyces cerevisiae*. **FEMS Microbiology Letters,** v.48, p.249-254,1987.

GADD, G. M. Accumulation of metals by microorganisms and algae. In: **Biotechnology: A Complete Treatise,** ed. REHM, H.; REED, G. v. 6b, p.401-433, 1988.

GADD, G. M. Biosorption. **Chemistry and Industry,** v.2, p.421-426, 1990.

GADD G. M; GHARIEB G. G. Role of glutathione in detoxification of metal by *Saccharomyces cerevisiae*. **BioMetals,** v. **17, p.**183-188, 2004.

GALUN, M.; KELLER, P.; MALKI, D. Removal of uranium (VI) from solution by fungal biomass and fungal wall related biopolymcrs. **Science,** v.219, p.285-286, 1983.

GAVRILESCU, M. Removal of heavy metals from the environment by biosorption.

Engineering in Life Sciences, v.4, n.3, p.219-231, 2004.

GHARIEB, M. M. Pattern of cadmium accumulation and essential cations during growth of cadmium-tolerant fungi. **Biometals,** v.14, p.143-151, 2001.

GOMES, D. S.; FRAGOSO, L. C.; RIGER, C. J.; PANEK, A. D.; ELEUTHERIO, E. C. S.

Regulation of cadmium uptake by *Saccharomyces cerevisiae*. **Biochimica et Biophysica Acta,** v.1573, p.21-25, 2002.

GROTEN, J. P.; BLADEREN, P. J. Cadmium bioavailability and healt risk in food. **Trends Food Sciences Techonogy,** v.5, p.50-55, 1994.

HANKS, D. L.; SUSSMAN, A. S. The relation between growth conidiation and trehalase activity in *Neurospora crassa*. **American Journal of Botany,** v.56, p.1160-1166, 1969.

HINO, A.; MIHARA, K.; NAKASIMA, K.; TAKANO, H. Trehalose levels and survival ratio of freeze-tolerant versus freeze-sensitive yeasts. **Applied and Environmental Microbiology,** v.56, p.1386-1391, 1990.

HOTTIGER, T.; BOLLER, T.; WIEMKEN, A. Rapid changes of heat and desiccation tolerance correlated with changes of trehalose content in *Saccharomyces cerevisiae* cells subjected to temperature shifts. **FEBS Letters,** v.220, p.113-115, 1987.

HUANG, C. P.; WESTMAN, D.; QUIRK, K.; HUANG, J. P. The removal of cadmium (II) from dilute aqueous solutions by fungal adsorbent. **Water Science Technology,** v.20, p.369 376, 1988a.

HUANG, C. P.; WESTMAN, D.; QUIRK, K.; HUANG, J. P.; MOREHART, A. L. Removal of cadmium (II) from dilute solutions by fungal biomass. **Particulate Science and Technology,** v.6, p.405- 419, 1988b.

HOUNSA, C. G.; BRANDT, E. V.; THEVELEIN, J.; HOHMANN, S.; PRIOR, B. A. Role of trehalose in survival of *Saccharomyces cerevisiae* under osmotic stress. **Microbiology,** v.144, p. 671-680, 1998.

HSDB HAZARDOUS SUBSTANCE DATA BANK. Cadmium. In: TOMES CPS ™

SYSTEM. **Toxicology, Occupational Medicine and Environmental Series,** Englewood: Micromedex, 2000. CD-ROM.

IKEDA, M.; ZHANG, Z. W.; MOON, C. S.; SHIMBO S.; WATANABE, N. H.; MATSUDA, N. I. Possible effects of environmental cadmium exposure on cadmium function in the

Japanese general population. **International Archives of Occupational and Environmental Health,** v.73, n.1, p.15-25, 2000.

ILO INTERNATIONAL LABOR OFFICE. **Encyclopaedia of Occupational Health and Safety,** Geneva, 1998. CD-ROM.

INTERNATIONAL ATOMIC ENERGY AGENCY. Vienna, 1987. **Comparison of Nuclear Analytical Methods with Competitive Methods** (IAES-TECDOC 435).

INTERNATIONAL ATOMIC ENERGY AGENCY. Vienna, 1990. **Practical Aspects of Operating a Neutron Activation Analysis Laboratory** (IAES-TECDOC 564).

INOUE, H.; SHIMODA, C. Changes in trehalose content and trehalase activity during spore germination in fission yeast *Schizosaccharomyces pombe*. **Archives of Microbiology,** v.129, p.19-22, 1981.

JAMIESON D. Saving sulfur. **Nature Genetics,** v 31, p 228-230, 2002.

JARUP, L.; BERLUND, M., ELINDER, C. G.; NORDBERG, G.; VAHTER, M. Health effects of cadmium exposure a review of literature and a risk estimate. **Scandinavian Journal of Work, Environment & Health,** v.24, p.1-51, 1998.

KAPLAN, D.; CHRISTIAEN, D.; ARAD, S. M. Chelating properties of extracellular polysaccharides from *Chlorella* app. **Applied Environmental Microbiology,** v.53, n.9, p.2953-2956, 1987.

KAPOOR, A.; VIRARAGHAVAN, T. Fungal biosorption - an alternative treatment. Option for heavy metal bearing wastewaters: a Review. **Bioresource Technology,** v.53 p.195-206, 1995.

KEFALA, M. L.; ZOUBOULIS, A. L.; MATIS, K. A Biosorption of cadmium ions by *Actinomycetes* and separation by flotation. **Environmental Pollution**, v.104, p.283-293, 1999.

KIFF, R. J.; LITTLE, D. R. Biosorpition of heavy metals by immobilized fungal biomass. In: **Immobilization of Ions by Biosorption,** ed. H. H. Eccles; S. Hunt. Ellis Horwood, Chichester, UK, p.71-80, 1986.

KURODA, K. & UEDA, M. Adsorption of cadmium ion by cell surface-engineered yeasts displaying metallothionein and hexa-His. **Applied Environmental Microbiology, v.**63 n.2 p.182-186.

KRUGER, P. **Principles of Activation Analysis**. New York: John Wiley, 1971.

KURTZMAN, C. P.; FELL, J. W. The yeasts: a taxonomic study. 4 ed. **Elsevier Science Publishers,** Amsterdam. 1998.

LANE, W. T.; SAITO, A. M.; GEORGE, N, G.; PICKERING, J. I.; PRINCE, C. R.; MOREL, M. M. F. Biochemistry: A cadmium enzyme from a marine diatom. **Nature**, v.435, p.42-42, 2005.

LASKEY, J. W.; REHNBERG G. L. Reproductive effects of low acut doses of cadmium chloride in adult male rats. **Toxicology and Applied Pharmacology**, v.73, p.250-255, 1984.

LESLIE, S. B.; ISRAELI, E.; LIGHTHART, B.; CROWE, J. H.; CROWE, L. M. Trehalose and sucrose protect both membranes and proteins in intact bacteria during drying. **Applied Environmental Microbiology**, v.61, p.3592-3597, 1995.

LEWIS, J. G.; LEARMONTH, R. P.; WATSON, K. Induction of heat, freezing and salt tolerance by heat and salt shock in *Saccharomyces cerevisiae*. **Microbiology**, v.141, p.687694, 1995.

LI Z.; LU Y.; ZHEN R.; SZCZYPKA M.; DENNIS J. THIELE AND PHILIP A. R. A new pathway for vacuolar cadmium sequestration in *Saccharomyces cerevisiae*: YCF1-catalyzed transport of bis (glutathionate) cadmium. **Proceedings of the National Academy of Sciences of the United States of America,** v. 94, p. 42-47, 1997.

LILLIE, S. H.; PRINGLE, J. R. Reserve carbohydrate metabolism in *Saccharomyces cerevisiae*: responses to nutrient limitation. **Journal Bacteriology,** v.143, p.1384-1394, 1980.

LINDQUIST, S. L.; CRAIG, E. A. The heat-shock proteins. **Annual Review of Genetics**, v.221, p.631-677, 1988.

LONDESBOROUGH, J.; VARIMO, K. Characterization of two trehalose in baker's yeast. **Biochemical Journal,** v.219, p.511-518, 1984.

LÓPEZ ALONSO, M.; BENEDITO, J. L.; MIRANDA, M.; CASTILLO, C.; HERNANDEZ, J.; SHORE, R. F. Arsenic, cadmium, lead, copper and zinc in cattle from Galicia, NW Spain. **Science of the Total Environment,** v.246, p.237-248, 2000.

LUCERO, P.; PENALVER, E.; MORENO, E.; LAGUNAS, R. Internal trehalose protects endocytosis from inhibition by ethanol in *Saccharomyces cerevisiae*. **Applied Environmental Microbiology,** v.66, n.10, p.4456-4461, 2000.

MACASKIE, L. E.; DEAN, A. C. R.; CHEETHAM, A. K. Cadmium accumulation by a *Citrobacter* ssp.; the chemical nature of the accumulated metal precipitate and its location on the bacterial cells. **Journal General Microbiology,** v.133, n.13, p.538-544, 1987.

MANSURE, J. J. C.; PANEK, A. D.; CROWE, L. M.; CROWE, J. H. Trehalose inhibits ethanol effects on intact yeast cells and liposomes. **Biochimica et Biophysica Acta,** v.1191, p.309-316, 1994.

MASON, J. O. **Toxicological profiles**: cadmiun.   Available at

http://www.atsdr.cdc.gov/interactionprofiles/IP-metals1/ip04-a.pdf . Accessed on June 13, 2005.

MATIS, K. A.; ZOUBOULIS, A. I. Flotation of cadmium: loaded biomass. **Biotechnology & Bioengineering,** v.44, n.3, p.354-360, 1994.

MATTIAZZO-PREZOTTO, M. M. **Behavior of copper, cadmium and zinc added to tropical soils at different pH values.** 197p. Thesis (Livre Docência) Escola Superior de Agricultura "Luiz de Queiroz", Universidade de Sao Paulo. Piracicaba, 1994.

McMURRAY, C. T.; TAINER, J. A. Cancer, cadmium and genome integrity. **Nature Genetics,** v.34, n.3, p.239-241, 2003.

MEDTEXT MEDICAL MANAGEMENT. Cadmium. In: TOMES CPS TM SYSTEM.

Toxicology, occupational medicine and environmental series. **Englewood: Micromedex,** 2000. CD-ROM.

MENDOZA-COZATL, D.; LOZA-TAVERA, H.; HÉRNÂNDEZ-NAVARRO, A.; MORENO-SANCHEZ, R. Sulfur assimilation and glutathione metabolism under cadmium stress in yeast, protists and plants. **FEMS Microbiology Review,** 2004.

MENEZES, M. A. B. C.; SABINO, C. V. S.; FRANCO, M. B.; KASTNER, G. F.; MONTOYA, E. H. R. K0-instrumental neutron activation establishment at CDTN, Brazil: a successful story. **Journal of Radioanalytical and Nuclear Chemistry,** v. 257, n.3, p. 627632, 2003.

MIDIO, A. F.; MARTINS, D. I. **Toxic agents, indirect food contaminants.** In: Toxicologia de alimentos. Sao Paulo: Varela, 2000. Chap. IV, p.163-252.

MIRANDA, C. E. S. **Determination of cadmium by atomic absorption spectrophotometry with preconcentration in ion exchange resin using an FIA system.** 89p. Master's degree dissertation - Sao Carlos Institute of Physics and Chemistry, University of Sao Paulo. Sao Carlos, 1993.

WATANABA M. Can bioremediation bounce back? **Nature Biotechnology, v.19, p.11111115, 2001.**

MORAIS, P. B.; ROSA, C. A.; LINARDI, V. R.; PATARO, C.; MAIA, A. B. R. A. Characterization and succession of yeast populations associated with spontaneous

fermentation during the production for Brazilian sugarcane aguardente. **World Journal of Microbiology & Biotechnology,** v.13, p.241-243, 1997.

MUZZARELLI, R. A. A., TANFANI, F.; SCARPINI, G. Chelating, film-forming, and coagulating ability of the chitosan-glucan complex from *Aspergillus niger* industrial wastes. **Biotechnology & Bioengineering**, v.22 ,p.8858-96, 1980.

NEVES, M. J.; JORGE, J. A.; FRANÇOIS, J. M.; TERENZI, H. F. Effects of heat shock on the level of trehalose and glycogen, and on the induction of thermotolerance in *Neurospora crassa*. **FEBS Letters**, v.283, p.19-22, 1991.

NEVES, M. J.; FRANÇOIS, J. On the mechanism by which a heat shock induces trehalose accumulation in *Saccharomyces cerevisiae*. **Biochemical Journal**, v.288, p.859-864, 1992.

NEVES, M. J.; TERENZI, H. F.; LEONE, F. A.; JORGE, J. A. Quantification of trehalose in biological sample with conidial trehalase from thermophilic fungus *Humicola grisea* var. *thermoidea*. **World Journal of Microbiology & Biotechnology**, v.10, p.17-19. 1994.

OLENA K. VATAMANIUK, ELIZABETH A. BUCHER, JAMES T. WARD & PHILIP A. R. A New Pathway for Heavy Metal Detoxification in Animals. **THE JOURNAL OF BIOLOGICAL CHEMISTRY** Vol. 276, N. 24, v.15, p. 20817-20820, 2001

OLSON G.J.; BRIERLEY JA.; BRIERLEY C.L. Bioleaching review part B: Progress in bioleaching: applications of microbial process by the minerals industries. **Applied Microbiology and Biotechnology**, v. 63, p. 249-257, 2003.

PALMISANO A. & HAZEN T. Bioremediation of Metals and Radionuclides: What It Is and How It Works (2nd Edition). **Lawrence Berkeley National Laboratory**, 2003. <http://repositories.cdlib.org/lbnl/LBNL-42595_2003/> Accessed on: August 19, 2005.

PANEK, A. D. Trehalose metabolism - new horizons in technological applications. Brazilian Journal of Medical and Biological Research, v.28. p.169-181, 1995.

PARROU, J. L.; JULES, M.; BELTRAN, G.; FRANÇOIS, J. Acid trehalase in yeasts and filamentous fungi: localization, regulation and physiological function. **FEMS Yeast Research**, v.5, n.6-7, p.503-511, 2005.

PATARO, C.; GUERRA, J. B.; PETRILLO-PEIXOTO, M. L.; MENDONÇA-HAGLER, L.C.; LINARDI, V. R. ROSA, C. A. Yeast communities and genetic polymorphism of *Saccharomyces cerevisiae* strain associated with artisanal fermentation in Brazil. **Journal of Applied Microbiology**, v.89, p.24-31, 2000.

PETTERSON, A.; KUNST, L.; BERGMAN, B.; ROOMAM, G. Accumulation of aluminium by *Anabaena cylindrica* into polyphosphate granules and cell walls and X-ray dispersive microanalysis study. **Journal General Microbiology**, v.131, p.2545-2548, 1985.

PRASAD, M. N. V. Cadmium toxicity and tolerance in vascular plants. **Environmental and Experimental Botany**, v.35, n.4, p.525-545, 1995.

RAGAN, H. A.; MAST, T. J. Cadmiun inhalation and male reproductive toxicity. **Reviews of Environmental Contamination and Toxicology**, v.114, p.1-22, 1990.

RAO, C. R. N.; IYENGAR, L.; VENKOBACHAR, C. Sorption of copper (II) from aqueous phase by waste biomass. **Journal Environmental Engineering Division,** Am. Soc. Civ. Engrs, v.119, p.369-377, 1993.

RIBEIRO, M. J. S.; LEÂO, L. S. C.; MORAIS, P. B.; ROSA, C. A.; PANEK, A. D.
Trehalose accumulation by tropical yeast strains to stress conditions. **Antonie van Leeuwenhoek,** v.75, p.245-251, 1999.

ROELS, H. A.; HOET, P; LISON, D. Usefuness of biomarkers of exposure to inorganic mercury, lead, or cadmium in controlling occupational and environmental risks of nephrotoxicity. **Renal Failure,** v.21, n.3-4, p.251-262, 1999.

ROMANDINI, P.; TALLANDINI, L.; BELTRAMINI, M.; SALVATO, B.; MANZANO, M. DE BERTOLDI, M.; ROCCO, G. P. Effects of copper and cadmium on growth, superoxide dismutase and catalase activities in different yeast strains. **Comparative Biochemistry and Physiology,** v.103, n.2, p.255-262, 1992.

ROSS, I. S.; TOWNSLEY, C. C. The uptake of heavy metals by filamentous fungi. In: **Immobilization of Ions by Biosorption.** ECCLES, H ; HUNT, S. Chichester: Ellis Horwood, 1986. p.49-57.

SAN MIGUEL, P. F.; ARGUELLES, J. C. Differential changes in the activity of cytosolic and vacuolar trehalases along the growth cycle of *Saccharomyces cerevisiae*. **Biochimica et Biophysica Acta,** v.1200 n.2 p.155-160, 1994.

SEBRAE. **Agro-industrial system of cachaça de alambique; Technical study of alternatives for using sugar cane.** BH: 2004. Available at:
<:http://www.sebraemg.com.br/Geral/ arquivo_get.aspx?cod_documento=15 -> Accessed on June 15. 2005.

SHIEAISHI, E.; INOUHE, M.; JOHO, M.; TOHOYAMA, H. The cadmium-resistant gene, CAD2, which is a mutated putative copper-transporter gene (PCA1), controls the intracellular cadmium level in the yeast *S. cerevisiae*. **Current Genetics,** v.37, p.79-86, 2000.

SIDERIUS, M.; VAN WUYTSWINKEL, O.; REIJENGA, K. A.; KELDERS, M.; MAGER, W. H. The control of intracellular glycerol in *Saccharomyces cerevisiae* influences osmotic stress response and resistance to increased temperature. **Molecular Microbiology,** v.36, n.6, p.1381-1390, 2000.

SIEGEL, S.; KELLER, P.; GALUN, M.; LEHR, H.; SIEGEL, B.; GALUN, E. Biosorption of lead and chromium by *Penicillium* preparations. **Water Air & Soil Pollution,** v.27, p.69-75, 1986.

SIEGEL, S. M.; GALUN, M.; KELLER, P., SIEGEL, B. Z.; GALUN, E. Fungal biosorption: a comparative study of metal uptake by *Penicillium* and *Cladosporium*. In: PATTERSON, J. W.; R. PASSINO. **Metal Speciation, Separation and Recovery,** Chelsea, Michigan: Lewis Publishers, 1987. p. 339-361.

SILVA, S. M. G.; **The effect of cadmium on alcoholic fermentation and the use of vinasse to mitigate its toxic action.** 142p. Thesis (Doctorate) - Escola Superior de Agricultura "Luiz de Queiroz", Universidade de Sao Paulo, Piracicaba, 2001.

SINGER, M. A.; LINDQUIST, S. Multiple effects of trehalose on protein folding *in vitro* and *in vivo*. **Molecular Cell,** v.1, p.639-648, 1998.

SPOSITO, G. The chemical forms of trace metals in soil. In: THORNTON, I. **Applied Environmental Geochemistry**, London: Academic Press, 1983. chap 3, p.123-170.

SUSSMAN, A. S.; LINGAPPA, B. T. Role of trehalose in ascospores *of Neurospora tetrasperma*. **Science,** v.130, p.1343. 1959.

THEVELEIN, J. M. Regulation of trehalose mobilization in fungi. **Microbiology Review,** v.48, p.42-59, 1984.

THEVELEIN, J. M. Regulation of trehalose metabolism and its relevance to cell growth end function. **The Mycota Biochemistry and Moleccular Biology**. v.3, p.395-420, 1996.

TOBIN, J. M.; COOPER, D. G.; NEUFELD, R. J. Uptake of metal ions by *Rhizopus arrhizus* biomass. **Applied Environmental Microbiology,** v.47, p.821-824, 1984.

TOWNSLCY, C. C.; ROSS, I. S.; ATKINS, A. S. Biorecovery of metallic residues from various industrial effluents using filamentous fungi. In: LAWRENCE, R. W.; BRANION, R.

M. R.; ENNER, H. G. **Fundamental and Applied Biohydrometallurgy**, Amsterdam: Elsevier, 1986. p. 279-289.

TSEZOS, M.; VOLESKY, B. Biosorption of uranium and thorium. **Biotechnology & Bioengineering**, v.23, p.583-604, 1981.

TYLER, G. Bryophites and heave metals: a literature review. **Botanic Journal of Linnean Society**, v.10, p.221-235, 1990.

TYNECKA, Z.; GOS, Z.; ZAJIE, J. Energy: dependent efflux of cadmium coded by a plasmid resistance determinant in *Staphylococcus aureus*. **Journal of Bacteriology**, v.147, n.7, p.313-319, 1981.

U.S. GEOLOGICAL SURVEY. **Mineral commodity summaries**. Available at: <http://geo-nsdi.er.usgs.gov/metadata/mineral/mcs/metadata.html>. Accessed on: June 13, 2005.

VAN AELST, L.; HOHMANN, S.; BULAYA, B.; DE KONING, W.; SIERKSTRA, L.; NEVES, M. J.; LUYTEN, K.; ALIJO, R.; RAMOS, J.; COCCETTI, P.; MARTEGANI, E.; MAGALHÂES-ROCHA, N. M.; BRANDÂO, R. L.; VAN DIJCK, P.; VANHALEWYN, M.; DURNEZ, P.; JANS, A. W. H.; THEVELEIN, J. M. Molecular cloning of gene involved in glucose sensing in the yeast *Saccharomyces cerevisiae*. **Molecular Microbiology,** v.8, n.5, p.43-927, 1993.

VAN DIJCK, P.; COLAVIZZA, D.; SMET, P.; THEVELEIN, J. M. Differential importance of trehalose in stress resistance in fermenting and nonfermenting *Saccharomyces cerevisiae* cells. **Applied Environmental Microbiology,** v.61, p.109-115, 1995.

VAN LAERE, A. Trehalose, reserve and/or stress metabolite? **FEMS Microbiology Review,** v.63, p.201-210, 1989.

VEGLIO, F.; BEOLCHINI, F. Removal of metals by biosorption: a review. **Hydrometallurgy,** v.44, p.301-316, 1997.

VIANA, C. R. **Effect of thermal and alcoholic stress on the metabolism of trehalose and the expression of heat shock proteins in** *Saccharomyces cerevisiae* **strains isolated during the production of cachaça from Minas Gerais**. 91p. Dissertation (master's degree) - Department of Microbiology, ICB, Federal University of Minas Gerais, 2003.

VOLESKY, B. Biosorption and biosorbents. In: **Biosorption of Heavy Hetals,** Boca Raton: CRC Press, chap.1.1, p.3-6, 1990a.

VOLESKY, B. Biosorption by fungal biomass. In: **Biosorption of Heavy Hetals,** Boca Raton: CRC Press, chap.2.3, p.139-171, 1990b.

VOLESKY, B. Removal and Recovery of heavy metals biosorption. In: **Biosorption of Heavy Hetals,** Boca Raton: CRC Press, chap.1.2, p.7-43, 1990c.

VOLESKY, B.; MAY-PHILLIPS, H. A. Biosorption of heavy metals by *Saccharomyces cerevisiae*. **Applied Microbiology and Biotechnology,** v.42, n.5, p.797-806, 1995.

VUORIO, O. E; KALKKINEN, N.; LONDESBOROUGH, J. Cloning of two related genes encoding the 56-kDa and 123-kDa subunits of trehalose synthase from the yeast *Saccharomyces cerevisiae*. **European Journal of Biochemistry,** v. 216, p.849-861, 1993.

WAIHUNG, L.; CHUA, H.; LAM, K. H.; BI, S. P. A comparative investigation on the biosorption of lead by filamentous fungal biomass. **Chemosphere,** v.39, p.2723-2736, 1999.

WALES, D. S.; SAGAR, B. F. Recovery of metal ions by microfungal filters. **Journal of Chemical Technology & Biotechnology,** v.49, p.345-355, 1990.

WINDERICKX, J.; DE WINDE, J. H.; CRAUNWELS, M.; HINO, A.; HOHMANN, S.;

VAN DIJCK, P.; THEVELEIN, J. M. Expression regulation of genes encoding subunits of the trehalose synthase complex in *Saccharomyces cerevisiae*. Novel variations of STRE- mediated transcriptional control? **Molecular Genetics and Genomics,** v.262, p.470-482, 1996.

WHO WORLD HEALTH ORGANIZATION. **Cadmium**. Geneva (Environmental Health Criteria 134) 1992.

WOOD, J. M.; WANG, H. K. Microbial resistance to heavy metals. **Environmental Science Technology,** v.17, n.12, p. 582-590, 1983.

YARROW, D. Methods for the isolation, maintenace, classification and identification of yeasts. *In* The yeasts: a toxonomic study. 4 ed. Edited by C.P. Kurtzman and J.W. Fell. **Elsevier Science B.V.** Amsterdam. p.77-100, 1998.

ZAVON, M. R.; MEADOW, C. D. Vascular sequelae to cadmium fume exposure. **American Industrial Hygiene Association Journal,** v.31, p.180-182, 1970.

ZUZUARREGUI, A.; DEL OLMO, M. Analyses of stress resistance under laboratory conditions constitute a suitable criterion for wine yeast selection. **Antonie Van Leeuwenhoek,** v.85, n.4, p.271-280, 2004.

Printed by Books on Demand GmbH, Norderstedt / Germany